Edward Robert Pearce Edgcumbe

Popular fallacies regarding bimetallism

Edward Robert Pearce Edgcumbe

Popular fallacies regarding bimetallism

ISBN/EAN: 9783337131425

Printed in Europe, USA, Canada, Australia, Japan

Cover: Foto ©berggeist007 / pixelio.de

More available books at **www.hansebooks.com**

POPULAR FALLACIES

REGARDING

BIMETALLISM

BY

Sir ROBERT P. EDGCUMBE

London
MACMILLAN AND CO., Ltd.
NEW YORK: THE MACMILLAN CO.
1896

All rights reserved

"I know nothing about 'the Currency Question'; isn't it something about—what's the word—bimetallism?"

"Yes. It's a pity you don't understand it. The number of those who don't is diminishing; and the number of those who do is increasing. The latter are called bimetallists and the former monometallists."   *"The Lesters,"* by Sir GEORGE CHESNEY.

"Every matter has two handles, one of which will bear taking hold of, the other not."   EPICTETUS.

PREFACE

This little treatise deals with a subject which has employed many pens, but deals with it in a new way. Adopting the method Bastiat followed in his Free Trade *Sophismes Économiques*, I have selected the leading current phrases, made use of by those who are opposed to the restoration of the joint standard of currency. These phrases embody the arguments monometallists rely upon, in their endeavour to maintain the existing monetary anarchy, and each of these arguments is examined in turn upon its merits.

In putting to the test these several monometallist fallacies, I have striven to be concise, to use language clear and simple to those who have not made a study of economic writings, and to base the principle of a common and a stable currency upon reason, rather than upon the passing conditions of trade and commerce.

Professor Foxwell has done me the great kindness to read over the proof-sheets, and while he cannot, under the conditions of a hurried revision, be held responsible for any errors which may remain, I wish most gratefully to acknowledge many valuable suggestions and corrections from his pen.

R. P. E.

SOMERLEIGH COURT, DORCHESTER,
    2nd November 1896.

CONTENTS

I

"A Monometallic Gold Currency is sufficient for our Needs" . . . . PAGE 1

II

"Bimetallism was abandoned by England for Sound Economic Reasons" 5

III

"France and other Countries abandoned Bimetallism by Choice and presumably for Good Reasons" . . . 15

IV

"The Relative Value of Gold and Silver cannot be fixed any more than the Relative Value of Cutlery and Broadcloth" 20

V

"If Gold had risen in Value all Prices would have fallen equally" 25

VI

"The Fall in the Price of Silver is due to the Excessive Output" 31

VII

"The Fall in Prices is accounted for by the Supply of Commodities outrunning the Demand, owing to Increased Production and better Facilities of Transport" . . 35

VIII

"Bimetallism is a Quack Remedy for Bad Times, and Bad Times must come now and again" . . . 42

IX

"A Common Currency for the World is not important, for all Trade is Barter" 45

X

"The United States are a Standing Warning against Bimetallism" 47

XI

"The Gresham Law (so-called) proves the Impracticability of Bimetallism" 57

XII

"Bimetallism spells Dishonest Money" . . 59

XIII

"As the Banks and Treasuries of the World are overflowing with Gold, it is absurd to argue that Gold is scarce" . 65

XIV

"Bimetallism is only a Nostrum to help Debtors" . . . 69

XV

"Low Prices are not an Evil, on the contrary an Advantage": The "Weights and Measures" Fallacy 71

XVI

"Bimetallism would depreciate the Currency, and to depreciate the Currency is to rob Labour" 78

XVII

"Savings Bank Returns, School, Criminal, and other Statistics prove that all is well" 86

XVIII

"Wanted a Ratio? Bimetallists have no Definite Plan" . 90

XIX

"Bimetallism would double Prices" . . 98

XX

"Bimetallism is Protection" . . . 102

XXI

"A Fixed Ratio cannot be maintained" . 106

XXII

  PAGE

"Any Alteration of an Existing Standard of Value is an Evil"   111

XXIII

"Bimetallism would be a Dangerous Experiment, and a throwing of the Sovereign into the Crucible" .   .   .   113

XXIV

"Bimetallism would make England a Dumping-Ground for Silver, and drive Gold out of Circulation"   .   .   117

XXV

"If the Mints were reopened to Silver, every one with a Balance would draw it out in Gold and every Bank would break"   120

XXVI

"The Supply of Gold coming forward from the Transvaal and elsewhere will solve the Question"   .   .   .   125

XXVII

"A Single Standard of Currency is better than a Double Standard"   128

XXVIII

"If Bimetallism raised the Price of Silver, the Gain of the Silver Users would be the Loss of the Gold Users"   .   130

XXIX

"Bimetallism means Inflation of the Currency"   133

XXX

"The Gold Standard is the Cause of England's Prosperity" .   137

XXXI

"Soft" Money and Paper Currencies : Effect on Wheat Prices   142

XXXII

Summary   146

APPENDIX—

    1. Sauerbeck's Index Numbers, 1820-1895 .   151
    2. Commercial Ratio of Silver to Gold, 1687-1895 .   152

INDEX   153

I

"A MONOMETALLIC GOLD CURRENCY IS SUFFICIENT FOR OUR NEEDS"

An Association has been formed called the Gold Standard Defence Association—"to oppose Bimetallism and unite in Defence of the Gold Standard" as "essential to the commercial position of our country and to the due discharge of contracts." "The treasuries of the world are overflowing with gold," so much so that it is "absurd to say that there is any scarcity of it." Bimetallism applied to "future contracts means confusion, and to existing contracts fraud and robbery." Bimetallism would "depreciate the currency." Bimetallists "refuse to answer what ratio should be fixed between gold and silver, though the question is a vital one." "It is a fatal objection to bimetallism that no one can foresee its ultimate result." Bimetallism is a "direct attack upon every artisan and every labourer." Bimetallists "propose to throw the English gold sovereign into the crucible." "The experience of eighty years has convinced us that the monometallic system is in every respect suited to our wants" and "that any serious attempt to modify it by the adoption of silver concurrently with gold (bimetallism) would be followed by consequences dangerous to the trade and commerce of this country."

Many other propositions are laid down by the Gold Standard Defence Association, but these are sufficient for the moment. The aim of this little treatise is to show that every one of these propositions involves a fallacy—that is to say, involves and conceals assumptions on matters of fact that are entirely erroneous.

The strength of the Gold Defence movement is no doubt due in part to interest, fortified by the fact that the monometallists are in possession —*beati possidentes;* but it rests still more upon error, upon incomplete truths.

Let me illustrate what I mean. It is common enough to hear said, "What do we want with bimetallism? Is not Gold good enough for us to do business with and to pay our way. Monometallism is good enough for me." And so should we all say had the world a gold currency. But the world has not a gold currency, and in similar plight is the British Empire, with gold currencies in some territories and silver currencies in others, linked by no fixed ratio of exchange. Yes. " Monometallism is good enough for me." But what do those who make use of this expression convey by it. They convey to those who have not exerted themselves to consider the matter carefully, the impression that gold " monometallism " provides a complete currency for the world. The term " monometallism " thus used is a question-begging expression. It contains a fallacy, and the strength of the fallacy lies in its setting forth an incomplete truth. The truth is visible to the external eye; the untruth can only be perceived by the mind.

Our monometallic friend sees that a currency based on a single metal (monometallism) is simpler and therefore better than a currency based upon two metals (bimetallism). We grant it, and so far he reasons justly. This is *that which is seen*. He does not perceive that the world does not possess a common metallic currency, and that even within the British Empire itself there exists no common currency, silver in some parts being current to the exclusion of gold, and gold in other parts current to the exclusion of silver. This is *that which is not seen*. He knows that when his wife orders in a supply, say, of rice or tea from the grocer's, she pays in gold. This is *that which is seen*. He does not perceive that the tea and the rice are purchased from the grower with silver and not with gold, and that how much silver the gold will exchange for varies from day to day, and that accordingly as between gold-using countries and silver-using countries there is no common medium of exchange. This is *that which is not seen*.

When therefore any one declares that he is in favour of monometallism as sufficient for our needs, he declares in favour of a currency which covers barely one-third of the world, instead of for a currency which is world-wide; he declares in favour of one-third of the population of the world carrying on trade with the other two-thirds of the world's population by the clumsy and elementary system of barter; he declares in short for a partial and imperfect currency instead of for one which is effective everywhere. This is *that which is not seen*.

Correct in his reasoning that a single standard is better than a joint standard, he does not perceive that a single standard does not exist, and is unattainable, and that in "opposing bimetallism" he commits himself to the untenable theory that it is better for the world to have no common standard of currency, than to have one based upon a joint standard. All this is *that which is not seen*.

A gold monometallist in search of a currency for the world's trade is pursuing a hopeless task—as hopeless, in the quaint illustration of Erasmus, as the search of a blind man in a dark room for a black cat that is not there.

II

"BIMETALLISM WAS ABANDONED BY ENGLAND FOR SOUND ECONOMIC REASONS"

"FRANCE abandoned bimetallism, and shall we be so very foolish as to take on the cast-off experiment of France?" asks the Gold Defence Association. England, France, and other countries have given up the joint standard, why adopt a policy those countries have cast aside? This method of stating the case enunciates a fallacy. *It is seen* that these countries have abandoned the joint standard, and the casual inquirer jumps to the fallacious conclusion that these countries, one by one, abandoned bimetallism of deliberate intent, after putting it to the test of experience and finding it wanting. *It is not seen* that the gradual abandonment of the joint standard was the result of a series of accidental causes and misconceptions, and not due to any inherent defect in the bimetallic system.

William III. found the English currency—silver was then legal tender all the world over—"in a state at which the boldest and most enlightened statesmen stood aghast" (Macaulay). Clipping and debasement of the coinage had been rife for a long period. To deal with the reformation of the currency three singularly able men were appointed—John Locke, Sir Isaac Newton, and Lord Somers; and in 1696 they established the English currency upon the bimetallic or joint-standard system. From careful inquiry made at that time they found that the relative value of silver to gold in England was as 15 to 1, and in neighbouring countries "about" 15 to 1. They decided that gold might be freely coined as well as silver, and the ratio they fixed upon was 1 oz. of gold to 16 (15.934) ozs. of silver—that is to say, the public were thenceforth entitled to deliver to the mint any quantity of silver, and were entitled to have it returned (the mint not charging for coinage) coined into silver at the rate of 4s. 11d. of silver coin for every ounce of silver. In a similar manner gold might be taken to the mint in any quantity, and the taker was entitled to receive back gold coin at the rate of £3 : 17 : 10½ for every ounce of gold tendered to the mint. Gold coins could only be obtained for gold bullion, and silver coins for silver bullion. Silver would not be accepted in payment for gold coin, nor gold bullion in payment for silver coin.

The ratio fixed in 1696 valued silver too low at 16 (15.934) to 1. It meant that guineas passed for 22s., and the result was that silver slipped away out of the country, and great inconvenience resulted. Three years later the ratio was accordingly altered to 15½ (15.572) to 1, making the guinea worth 21s. 6d., and raising the ounce of silver to 5s. 0½d. But still inconvenience was felt, for this ratio continued to value silver lower than in France, and silver tended to go where it possessed the greater purchasing value. So the English ratio was again altered in 1717 and fixed at 15½ (15.209) to 1, making the guinea worth 21s., and the ounce of silver worth 5s. 1½d.

When this final alteration was made Sir Isaac Newton recommended that our ratio should be made identical with that of France, but his advice was not followed and the English ratio continued through almost the whole century slightly more favourable to gold than was the French ratio. What effect this difference in the ratio had upon the English currency we shall see. This final readjustment in 1717 of the ratio continued the English currency upon a bimetallic basis of silver and gold down to the year 1816. No one can say that the eighteenth century covered a period of financial disaster, when it is remembered that at the close of the great war of the Austrian Succession (1740-1748) the English 3 per cent consols stood at par. Alison, writing of the trade of Great Britain before the English Mint was closed to silver, said, "the monopoly of almost all the trade of the world was in our hands."

Since 1816 the joint standard has been abandoned, and gold only has been coined for the public at the English Mint. "Yes, but we get our silver from the mint too," says one. We do, but we cannot purchase silver at the mint with silver, it now has to be paid for in gold, which makes the whole difference. The mint authorities purchase silver bullion as and when they require it. They are no longer under an obligation to receive it from any one and return it coined into money. Nor is any one entitled by law to tender in payment of a debt more than 40s. in silver. Since 1816 silver has circulated like copper money, bank notes (other than Bank of England notes, which are privileged) and cheques. It has become token money. Gold since that date has taken the place formerly occupied by both silver and gold. Silver since 1816 has been demonetised, that is, it has ceased to be lawful tender in payment of debts. It ranks with paper money (other than Bank of England notes), which the creditor is entitled to refuse and demand payment of his debt in gold.

Those who "oppose bimetallism" say that the free coinage of silver was abandoned in 1816 owing to the sagacity of Lord Liverpool, who, in conjunction with other eminent financiers, deemed it better for this country that its standard of currency should consist of gold and gold only. This theory of the sagacity of Lord Liverpool is unfortunately one which will not bear investigation. In the first place Lord Liverpool was in no degree an eminent man, he has indeed been not inaccurately termed by Lord Beaconsfield "the arch-mediocrity of English politics." In the second place, when we come to inquire what operated upon his mind to cause him to tamper with the joint standard established in 1696, it becomes abundantly clear, from the reasons he has left on record, that he was labouring under more than one serious misconception.

The first and chief reason which led Lord Liverpool to advocate this change was, that all through the eighteenth century the English Mint was mainly employed in coining gold and was little called on to coin silver. England is evidently a gold country, said Lord Liverpool, and there is nothing to be gained by continuing to coin silver that may be brought to the mint. Now, if Lord Liverpool had been the sagacious person some would have us believe, he would have quickly discovered, what had been pointed out again and again by expert writers, why it was that so little silver was coined during the eighteenth century at the English Mint. The reason was that the relative value of silver and gold differed slightly at the English and French Mints, it being more profitable to coin silver at the French Mint than at the English. The English ratio, fixed in 1696 at 16 of silver to 1 of gold, had been altered in 1717 to $15\frac{1}{2}$ to 1, and so continued until 1816. In France, from 1713 down to 1726, the silver ratio was slightly less than this, and from 1726 to 1785 was about $14\frac{1}{2}$ to 1. The consequence was that those who owed money to Frenchmen could pay their debts in that country either with an ounce of gold or $14\frac{1}{2}$ ozs. of silver, while in England they could pay Englishmen with an ounce of gold or $15\frac{1}{2}$ ozs. of silver. Naturally silver tended to go to France in payment of debts and gold to England. In the same way, in adjusting debts as between England and France, the Englishman sent silver in payment to France, where he needed only $14\frac{1}{2}$ ozs. as compared with $15\frac{1}{2}$ ozs. to make a similar payment in England. For the same reason a Frenchman remitted gold to England in payment and kept his silver. During the eighteenth century there were constant complaints of the want of small silver change in England, and the whole of the movement of silver to France and to the French Mint, and of gold to England and to the English Mint, was due to the divergent relative values (or ratios) of the two metals at the two mints. In 1785 the French Mint readjusted their ratio and brought it up to $15\frac{1}{2}$ to 1, when the movement of silver to France and of gold to London slackened immediately.

Why was it that during nearly the whole of this century, in spite of the inconvenience arising from want of silver change, the English ratio was not altered and we awaited readjustment by France? The answer is, that the English bankers preferred handling gold to handling silver. In the last century especially, when the means of communication were primitive as compared with the facilities of railway communication in the present day, it mattered a good deal to bankers whether they had to move gold or silver about the country. The public wanted silver change, but it gave the bankers less trouble to provide gold, so the public might want.

Adam Smith pointed out that there were two ways of getting over the difficulty of the want of small silver change, either by making the English ratio identical with the French ratio, or by demonetising silver, and converting it into token money. But there was a third way and a better way which Adam Smith did not perceive, and that was to have *both* standard and token coins of silver. That is to say the crown piece in silver might have been legal tender, each crown piece containing one ounce of pure silver, while the smaller silver coins might have been minted with less than their nominal relative amount of silver. The half-crown, for example, instead of containing half-an-ounce of silver, might contain only one-third of an ounce of silver, so would remain in circulation, being only of token value. This third way of dealing with silver was rightly adopted by the United States in 1853, when Congress enacted that all fractional silver coins under one dollar should be made of light weight. France also adopted the same principle in 1864. By this method every difficulty was surmounted. Silver on the one hand remained legal tender. On the other hand the smaller silver coinage was not liable to be withdrawn from the currency.

If Lord Liverpool had been the great authority we are asked to believe him to be, he would have perceived this and understood the operations we have explained. The only excuse that can be made for him is, that about the time France adjusted her ratio to the English ratio, the excessive note issues of numerous small banks throughout the country had greatly disturbed the currency, causing the monetary crisis of 1793, and following upon these sources of currency disturbance came the war with France which compelled us to resort to a forced paper issue (1797), and sent the precious metals to a premium.

Here we have, in a few words, the simple and incontrovertible causes of the abundance of gold currency and the deficiency of silver currency in England during the greater portion of the eighteenth century. But instead of perceiving the real causes at work, Lord Liverpool merely stumbled to the conclusion that gold was the "natural currency of England." He perceived how essential it was to increase the supply of small silver, but it never occurred to him that a silver token currency was compatible with the use of silver as legal tender. Had Lord Liverpool grasped this principle we should never have heard of monometallism or of bimetallism. Having come to the conclusion that gold "was the natural currency of England," but that small silver change was essential, he easily proceeded to the next assumption, that it would be desirable to demonetise the "unnatural" currency, namely silver.

We shall refer later to a parallel illustration of the effect of diverging ratios, as between the United States and France in the following century (X).

When Lord Liverpool committed himself to the principle that the maintenance of an open mint for silver was of no importance to England, it was suggested to him that if England closed her mint to the free coinage of silver, it was possible other countries might do the same and what then would the world do for a common currency if no countries retained a joint standard at a fixed ratio? To this Lord Liverpool made reply that other countries would not follow our example. Here Lord Liverpool fell into his second cardinal error. He was right indeed for a time, but only for a time. From 1816 down to 1873 France and the countries of the Latin Union—Italy, Belgium, Switzerland, and Greece—kept open their mints for the free coinage of silver and gold at the ratio of $15\frac{1}{2}$ to 1. So long as the Latin Union kept open their mints to both metals, the world suffered no ill effects from Lord Liverpool's tremendous change in the currency policy of this country, for the whole world retained its joint standard through the channel of the great Exchange kept open in Europe at the mints of the Latin Union. Until 1873 the Latin Union provided the world with a joint standard of currency, and the relative values of gold and silver remained for all practical purposes stationary about 1 to $15\frac{1}{2}$.

Two causes conspired to cloak the demonetisation of silver in 1816 and the effect it was to have upon the world. In the first place the change was introduced at a time (1816-1819) when England was exhausted by her long wars with France (1793-1815), and when, the currency having become depreciated by a forced paper issue with the precious metals standing at a premium, men's minds were intent upon getting the depreciated forced paper back to a par value, rather than upon considering what might be the possible ultimate effect of closing the mint to silver. In the second place the fact that on the continent the precedent set by us in 1816 was not followed up for more than half a century (1873), lulled men into complete disregard of the far-reaching change which Lord Liverpool, with his want of perspicacity, had brought about.

An attempt has been made to make Sir Robert Peel in part responsible for this change. Sir Robert Peel however was in 1816 a young statesman (twenty-eight years of age) immersed in Irish duties as Chief Secretary, and did not reach the cabinet until 1821, when the question had been dealt with. He was eager for the resumption of cash payments but did not trouble himself about the demonetisation of silver. When Sir Robert Peel in later years had to consider certain questions in relation to the currency, the question of the need of a joint standard as a medium of exchange for the settlement of international commercial transactions was not one on which he was called upon to decide, for the Latin Union was doing the world's service in this respect, and bridging the gulf between gold-using and silver-using countries with the joint standard. Yet in spite of this it appears that Sir Robert Peel had a clear perception of the obligations England was under to the Latin Union, for in a speech in the House of Commons (24th May 1844) he refers on the one hand to the inconvenience of our standard differing from other countries—gold in England, silver in Asia—and on the other hand to the use of silver for remittances as "answering all the purposes of gold"—which it could only do because, as he showed, it might always be sent (then) to Paris, where it was coined at the ratio of $15\frac{1}{2}$ to 1 of gold.

This brings us to the close of an important period. England having fixed her currency in 1696 on the basis of a joint standard, discarded it in 1816 for a single standard, and discarded it (*a*) in error as to gold being her "natural currency," (*b*) from not perceiving that it was possible to retain small silver in circulation by rendering a portion of the silver coins token money, and (*c*) in error as to other nations not following her precedent. The ultimate effect of the action of Lord Liverpool was not seen. Yet the whole question was to hinge upon the point, "Was the step taken by England to remain an isolated act, or was the abandonment of the joint standard to become a precedent for other countries." War and the consequent disturbances to finance, provided Lord Liverpool with an excuse for meddling with the currency. War and the financial trouble consequent thereon, compelled France unwillingly to withdraw from the Latin Union, and break up the compact with her neighbours for the free mintage of silver.

III

"FRANCE AND OTHER COUNTRIES ABANDONED BIMETALLISM BY CHOICE AND PRESUMABLY FOR GOOD REASONS"

DURING the year immediately following the increased output of gold from California and Australia (1850-1860) there were perceptible disturbances in the Exchange at the mercantile centres more particularly in touch with England, such as Bordeaux and Hamburg.

In France Napoleon III. supported the demand of the merchants of Bordeaux and urged the adoption of the gold standard, but the Bank of France and the financiers who were not mere theorists, perceived that these disturbances were at best only ripples on the surface, and that demonetisation of silver would cause far different and lasting effects.

In Germany the monometallist movement originated in a similar manner with merchants, especially at Hamburg, dealing largely with England. Feeling the inconveniences arising from slight fluctuations in the gold and silver exchange—difficulties described by Baron Alphonse Rothschild as "bien faibles"— the German merchants entirely ignored all consideration of the greater disturbances which would result from general demonetisation of silver, disturbances foreseen by the Bank of France, who in 1870 asserted in the strongest manner that they were perfectly satisfied with the working of bimetallism.

Thus it was that shortly before the Franco-German War in 1870, German statesmen had been considering what to do with their currency, which down to that time had been a silver currency only. Looking to England as a prosperous country, the German statesmen persuaded themselves that England's commercial position was due to her gold standard. As Lord Beaconsfield justly observed, " It is the greatest delusion in the world to attribute the commercial preponderance and prosperity of England to our having a gold standard. Our gold standard is not the cause of our commercial prosperity but the consequence of it" (Glasgow, November 1873). The fine diamonds a rich lady may wear are the consequence, not the cause of her wealth. But German statesmen reasoned otherwise. At that time Germany had no means of procuring gold in sufficient amount to furnish a new currency, but the war with France was at hand, and the issue of that conflict left Germany the creditor of France. The opportunity for placing the silver currency of Germany upon a gold basis had come, and payment of the indemnity of £200,000,000 was demanded in gold. But Germany not only demanded the payment of the war indemnity in gold—closing her own mints to silver in 1872—but was proceeding to pour her demonetised silver into France to be coined at the French mint into francs, and then exchanged in the market-place for gold pieces to be remitted to Germany. In 1872 the French mint coined 1,000,000 francs, but in 1873 was called on to coin 173,000,000 francs, and no less a sum than £80,000,000 of German silver was in sight ready to be poured into France for coinage purposes. France, alarmed at the prospect of losing all her gold, closed her mint to the free coinage of silver in self-defence, and the other countries of the Latin Union were immediately compelled to follow suit. In the same year (1873) the United States, at the bidding of the Senate, also closed the States mint to the free coinage of silver—but of this later. From this time the unprescient policy of Lord Liverpool began to take real effect.

It is worth observing that Soetbeer, who, as the mouthpiece of the Hamburg Chamber of Commerce, had more than any one else to do with the demonetisation of silver by Germany, became so alarmed at the consequences of his action, that he earnestly protested almost with his last breath (1892) against any extension of gold monometallism. The Gold Defence Association seem to think that they can dispose of the question, "Why did France abandon bimetallism?" by quoting the preamble to the French Currency Act of 1875, which, after a reference to the currency of Great Britain, goes on to say, "it may be stated that since 1857 the principle of the gold standard has won increasing favour"; but this deceives no one acquainted with the true position. This preamble was a mere gilding of the pill. A proud nation driven to close her mints to silver would hardly set forth in a "preamble" that the pressure of a war indemnity and causes resulting therefrom compelled her reluctantly to close her mints. If any weight could be attached to the "preamble" to the French Currency Act, France should be of the same opinion still, but after twenty years' experience of gold monometallism the French Government have declared that the demonetisation of silver was a mistake, and have expressed their desire to co-operate in the restoration of the joint standard. None can seriously doubt that the step taken by France in 1873 was a political and not a financial step, and that it was taken to counteract the movement of silver from her enemy Germany.

This view of the action of France in 1873 is generally confirmed by Mr. Bagehot, who, writing in July 1876, said :—" The Latin Union allowed any one to mint any quantity of either metal. In consequence they were always coining the metal of lesser value, and that metal was used to buy and take away the metal of higher value. In this way, in the cotton famine, France was half emptied of silver wanted for the East and filled with gold. If the Latin Union had adhered to this principle the great effect on the silver market produced by the German operations would have been rendered scarcely observable. As soon as silver began to fall it would have gone to France and been used to buy gold which had risen. Thus silver would have been taken from the general market and gold brought into it until the former level of values, or something like it, had been reached. But France and the Latin Union could not endure this. Partly from political and partly from economical motives, they would not take the 'cast-off' German silver." In short, France would not fetch and carry for Germany.

This completes the tale of the world's loss of a

common currency. To sum up. Prior to 1696 the currency of the world was on a silver basis. In that year England established a joint standard with a free mint for both silver and gold at the ratio, first of 16 and subsequently of $15\frac{1}{2}$ of silver to 1 of gold (see II.). In 1816 England closed her mints to silver, but this had no effect upon her trade, as France kept open her mints to both metals at the old ratio. In 1872 Germany, hitherto a silver-using country only, adopted gold monometallism, and in 1873 France, under pressure, and as a counter-move to Germany, closed her mints to silver. The other countries of the Union were compelled to do likewise. The United States, for reasons explained later, adopted the same course, and from that time, and for the first time, the world lost its common currency. The great exchange offices were closed. Gold and silver could no longer be exchanged at a fixed rate, and, as between gold-using and silver-using countries, the world was thrown back upon the elementary and barbaric methods of trade by barter.

IV

"THE RELATIVE VALUE OF GOLD AND SILVER CANNOT BE FIXED ANY MORE THAN THE RELATIVE VALUE OF CUTLERY AND BROADCLOTH"

"GOD preserve us from the evil spirit and from metaphors," said Paul Louis. How are men misled on these subjects. A few words misused and the mischief is done. How common it is to hear it said, even by men of considerable position in the State, "It is as impossible to fix the relative value of gold and silver as it is the relative value of bread and cheese, or of cutlery and broadcloth." This is a fallacy. *It is seen* that the relative value of commodities—such as bread and cheese, cutlery and broadcloth—cannot be regulated by statute. *It is not seen* that gold and silver are sometimes commodities and sometimes not. *It is not seen* that when gold and silver are employed as "money," that is, when used as "a standard of value" created by statute, they are not commodities. Nature produces the commodities gold and silver. Nature does not produce "money." It is the legislature which makes money—*Nomisma* (Gk. money) comes from *Nomos*

(Gk. law). *It is not seen* that when gold and silver are used for currency purposes, the large amount so required controls the value of gold and silver passing as commodities. Those who glibly talk of its not being possible to fix the relative value of gold and silver any more than the relative value of cutlery and broadcloth use metaphorical language, and the metaphor misleads them. These amiable persons are carried away by a fallacy and do not know it. John Locke, writing in 1691 a letter to a member of Parliament, said, "Suppose 15 to 1 be now the exact par between gold and silver, what law can make it lasting?" But five years later, in 1696, he modified his opinion and agreed with Sir Isaac Newton and Lord Somers in recommending a joint standard of currency with gold rated to silver at $16\frac{1}{2}$ to 1.

So long as the legislatures fixed the relative value of coined gold and silver, the relative value of the two metals as commodities remained practically stationary at the same relative value, the reason being that so large a bulk of the precious metals was needed for money, and was so easily converted into money at the mints, that no one was ready to sell either the one or the other, as a commodity, at more than a fraction below what could be obtained for it by converting it into currency. The mint price of the metals as money was accordingly reflected upon the uncoined metals when viewed as commodities, and as this was so in the past there is no reason to doubt that it would be so in the future.

To those who are not familiar with the theory of value, there might appear to be much weight at first sight, and apart from experience, in the theory that a fixed ratio cannot be maintained. As, however, we have the experience of nearly two centuries to guide us (1696-1873), any force such an argument might possess under other circumstances is lost. For not only did the fixed ratio exist unbroken from 1696 to 1873, but it remained throughout that period practically unchanged.

If it were impossible to fix the relative value of gold and silver money at a given ratio, it would have been made manifest during this century; for from 1830 to 1840 the annual output of gold was about £2,750,000, while the annual output of silver, taken at $15\frac{1}{2}$ ozs. of silver to 1 oz. of gold, was about £5,000,000. In 1853-54, owing to the increased output of gold in Australia and California, the annual output of gold had risen to £26,000,000,—nearly ten times its former output—while the annual output of silver had only crept up to £7,000,000. Yet this sudden and immensely increased output of gold acted only like a ripple on the parity of exchange, and the trade and commerce of the world went on as before. After such an experience it needs some hardihood to argue that a return to a fixed ratio is an impossibility.

The annual output of gold and silver bears so small a relative proportion to the existing stock of the two metals used for currency that it cannot appreciably affect them. It is legislation that maintains the value of the one or the other. Let the mines issue more silver money than gold or more gold money than silver, the public need not be startled by it nor the ratio disturbed, nobody ever being willing to part either with his silver or his gold at less than the legislative ratio.

The steadiness of the fixed ratio over nearly two centuries is so absolutely beyond dispute that the Royal Commission of 1886, composed of both monometallists and bimetallists, entirely agreed as to the practicability of the fixed ratio, and unanimously reported (No. 192) that, "So long as the bi-metallic system of the Latin Union was in force we think that, notwithstanding the changes in the production and use of the precious metals, it kept the market price of silver approximately steady at the ratio fixed by law between them, namely, $15\frac{1}{2}$ to 1." Again, "The fact that the owner of silver could in the last resort take it to the mints of the Latin Union and have it converted into coin which would purchase commodities at the ratio of $15\frac{1}{2}$ of silver to 1 of gold, would in our opinion be likely to affect the price of silver in the market generally, whoever the purchaser and for whatever country it was destined. It would enable the seller to stand out for a price approximating to the legal coin and would tend to keep the market steady at about that point."

So long as the Latin Union kept open their mints to both metals, every country, whether the standard was gold only, as in England, or silver only, as in India, stood in exactly the same position as if the mints were open to both metals. So that although in name England had the single gold standard and India the single silver standard, they enjoyed and in effect had the joint standard.

Professor Stanley Jevons, writing before the joint standard was abandoned in 1873, illustrated the operation of the joint standard in a very simple manner. Imagine two tanks, one fed with gold and one fed with silver. In the absence of any connecting pipe, the level of the water in each reservoir will vary with the volume of the supply or the quantity withdrawn. But if you open a connection, the water in both will assume a mean level, and the effect of a more abundant supply to one will be distributed over both reservoirs.

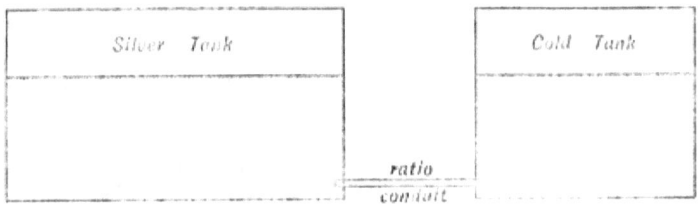

The mass of the metals, gold and silver, circulating in Western Europe in late years, is exactly represented by the water in these reservoirs, and the connecting point is the French law of 1803, which enables one metal to take the place of the other as an unlimited legal tender.

This illustration seems to have been in the mind of Mr. Bagehot when he wrote, " A very great part of the world adhered to the bimetallic system, which made both gold and silver legal tender, and established a fixed ratio between them. In consequence, whenever the value of the two metals altered, these countries acted as equalising machines. They took the metal which fell; they sold the metal which rose; and thus the relative value of the two was kept at its old point" (September 1876).

V

"IF GOLD HAD RISEN IN VALUE ALL PRICES WOULD HAVE FALLEN EQUALLY"

THE fallacy that "if gold had risen in value it would necessarily follow that the prices of all products would fall equally," has not only received the approval of the Gold Defence Association, but of a Chancellor of the Exchequer (17th March 1896). Sir Michael Hicks Beach propounded this fallacy to the House of Commons almost as if it were a new discovery which conclusively settled the question in favour of a single gold standard. Yet this fallacy was fully exposed more than a generation ago, before bimetallism was thought about, when the great gold discoveries of Australia and California affected prices by raising them. It was then stoutly argued by many that these discoveries could not have affected prices, because if so prices would have risen all round in the same ratio. Professor Cairnes devoted his powerful intellect to the analysis of this question, and proved conclusively that the operation of such a cause as the greater production of gold would affect prices of different commodities at different rates of speed and in different degrees, and that we must trace out the conditions of production of each class of commodities before we can discover how a greater or a less production of gold would act on prices. When one commodity becomes cheaper there is more money to spend upon other commodities. When the Corn Laws were repealed and bread became cheaper, meat and butter were more freely purchased, so cheap bread raised the price of meat and butter. The reasoning of Professor Cairnes is as good now to prove that if gold has appreciated, instead of, as then, depreciated, such appreciation would not affect all prices alike in the same degree, just as its depreciation did not affect all prices equally then.

Let us clear our minds of another prevalent fallacy—the fallacy that gold does not change in value because its price is fixed by law. Some who profess to be authorities, lay it down in the most decided way that the value of gold has not changed, and give as a reason that the price of gold is unchanged. "An ounce realises exactly the same price, namely £3 : 17 : 10½, as it did half a century ago. How then can the value of gold increase while its price remains unaltered?" But this constancy in price only shows that the quantity of gold in a sovereign remains the same. The mint authorities give £3 : 17 : 10½ for an ounce of gold, because they know that there is just sufficient gold in an ounce to manufacture three sovereigns, and that portion of a sovereign which is represented by 17s. 10½d. This is all that is meant. In short, that the price of gold remains unaltered merely means that the sovereign is of a fixed weight—rather more than a quarter of an ounce—and not that, as measured in other commodities, its value is constant. This is all that Sir Robert Peel meant when he asked his well-known question, "What is a pound?" and he might, if he had given its historical meaning, have stated that it originally meant a "pound-weight of silver." The old English pound was of 12 ozs. of silver, which at 5s. an ounce represented £3 of modern money. The modern penny was originally a pennyweight of silver—240 penny-weights making 1 lb. The silver pound and the silver penny of the Edwards came down by clipping and debasement to one-third of their value.

As we have now freed our minds from the fallacy that gold does not vary in value, because measured by itself its value is fixed, let us consider whether there are any clear indications that gold, when measured by other commodities, has risen in value.

As to one fact there can be no dispute and no doubt, and that is that since the closing of the mints by the Latin Union the relative value of gold and silver has altered. Before that happened the price of an ounce of silver as measured in gold was 5s. Now it is 2s. 7d., and has been as low as 2s. 3d. In coming to a decision as to which of the two metals has remained the more stationary in value, it seems clear that the metal which to-day retains most nearly its old purchasing power must be the metal which has changed least in value. That is to say if an ounce of gold exchanges to-day for very nearly the same quantities of various commodities as it exchanged for a generation ago, while an ounce of silver exchanges to-day for half as much of various commodities as it exchanged for thirty years ago, then silver has fallen in value while gold has remained stationary. But this is exactly what has not happened. An ounce of silver to-day exchanges for nearly as much land, wheat, tea, tin, and other commodities as it did thirty years ago; but an ounce of gold exchanges for nearly double the amount of commodities it did thirty years ago. As the value of commodities in the aggregate has followed the value of silver rather than the value of gold, can it be said that gold has remained stationary in value, while silver and commodities generally have fallen? As well might the Irishman on the jury who differed from his fellow jurymen complain that he "never met eleven more obstinate men in his life." The saddle must be put upon the right horse. It is evident that gold has risen largely in value while silver has at most but slightly fallen.

The following table, compiled by Mr. Sauerbeck, shows the relative purchasing power of silver and gold of recent years. The "index-number" is calculated on the wholesale price of forty-five commodities — namely, bacon, barley, beef (prime and middling), butter, coffee, cotton (American and Indian), copper, coal (London and export), flax, flour, hemp, hides, indigo, iron (bars and pig), jute, leather, lead, maize, mutton (prime and middling), oats, oil (olive, seed, and palm), potatoes, pork, petroleum, rice, soda (crystals and nitrate), silk, sugar (West Indian and Java), tallow, timber, tin, tea, wheat (English and American), wool (English and merino). In the table given the value of gold is taken as constant at 100. In the second column is shown the average yearly value of the above-mentioned forty-five commodities measured in gold.

In the third column is given the average yearly value of silver measured in gold, and in the fourth column the average yearly value of an ounce of silver.

TABLE I

SAUERBECK'S TABLES

(Set out fully in Appendix)

Gold.		Forty-five Commodities measured in Gold.	Silver measured in Gold.	Annual Average Price of Silver per Ounce.
1870	100	96	99	60¼d.
1875	100	96	93	56¾d.
1880	100	88	85	52½d.
1885	100	72	79	48½d.
1890	100	72	78	47¼d.
1895	100	62	49	29¾d.

This table shows that the price of commodities fell with the fall in silver, in other words that gold appreciated in value. The remarkable thing is, not that gold rose in purchasing power, but that silver retained in so large a degree its value, and in fact, until the closing of the Indian mints, practically showed no fall. When we consider that Germany, with a silver standard only, suddenly abandoned it for a gold standard, and that almost simultaneously the mints were shut to silver in France, Italy, Belgium, Switzerland, Greece, and the United States, the remarkable thing is, that silver has shown such extraordinary stability as a purchasing medium. This stability of silver is incompatible with the notion of the stability of gold entertained by the gold monometallists.

The fallacy that gold has not changed in value owes its vitality to the fact that, gold being the measure of value, we speak of prices of other commodities as declining when they diminish in relation to gold. Just in the same way we speak of the sun rising in the morning and setting at night, although all the time it is the world which is moving and not the sun.

The fallacy that gold has not risen in value assumes different guises—one form of it being met with in the remarkable statement that "the prices of produce have not fallen." This form of the fallacy has been set forth in the pages of *The Statist* (1895) and *The Economist* (1896) by Mr. Power, who has published statistics in these papers to prove that the produce prices of the Mississippi Valley have not fallen as measured in gold. Mr. Power takes the yield of acres in the Mississippi Valley in tons, lumping wheat, maize, oats, and potatoes together, and because this composite ton is worth more now than it was in the "seventies," therefore, says he, gold has not appreciated, on the contrary it has rather depreciated. It would be just as easy by such a method to show that in Sussex, for example, gold had depreciated in purchasing power of recent years rather than appreciated. We walk into a hop garden which ten years ago was a hayfield. As a hayfield it produced four tons of hay from nine acres, worth, £16. To-day it produces seven tons of hops, worth £100. The money value of the produce of a Sussex holding has risen 600 per cent. Therefore, according to this strange method of reasoning, gold has not appreciated in value, but on the contrary depreciated.

VI

"THE FALL IN THE PRICE OF SILVER IS DUE TO THE EXCESSIVE OUTPUT"

ONE of the most widely-accepted misconceptions as to the change in the relative value of silver and gold is, that silver has fallen in consequence of an excessive output from the mines. The most superficial inquiry is sufficient to show that this is only one of the many fallacies which have to do duty in default of good argument against a joint standard of silver and gold. It is quite true that after the closing of the mints of the Latin Union there was an increased output of silver from the silver mines. This is *that which is seen.* Those who have not made themselves conversant with the maintenance of a fixed ratio between gold and silver prior to 1873, jump to the conclusion that the increased output of silver upset the relative value and that this explains everything. Now, in the first place, it is clear that an increased output of one of the two metals must be beyond anything we have ever experienced, if, judging by past experience, it is to upset the fixed ratio (see IV.). This is *that which is not seen.* In the second place, the increased output of silver has in no degree been excessive and cannot account for the existing relative value of gold and silver. This is known to those who have studied the returns, but to the general public it is *that which is not seen*.

Let us examine the following table:—

TABLE II

RELATIVE PRODUCTION OF GOLD AND SILVER

For 100 years,	1681 to 1780,	26.3 ozs. of Silver to 1 oz. of Gold.
,, 50 ,,	1781 to 1830,	46.2 ,, ,,
,, 50 ,,	1831 to 1880,	8.7 ,, ,,
,, 5 ,,	1881 to 1885,	17.2 ,, ,,
,, 5 ,,	1886 to 1890,	20.9 ,, ,,
,, 5 ,,	1891 to 1895,	19.5 ,, ,,
,, 1 year,	1895,	17.3 ,, ,,

From this table it is seen that during half a century, 1781-1830, the output of silver to gold was nearly treble the ratio, viz. 46 to 1 instead of $15\frac{1}{2}$ to 1, yet during the whole of that period the ratio was maintained without the least difficulty. Again, during the next fifty years the output of silver fell considerably below the fixed ratio, viz. $8\frac{3}{4}$ ozs. of silver to 1 of gold instead of $15\frac{1}{2}$ ozs. Yet again there was no difficulty whatever in maintaining the ratio. Only after 1881 does the output of silver again begin to exceed the ratio of $15\frac{1}{2}$, and then, as the table shows, to a degree bearing no comparison with the excess of gold to silver from 1831 to 1880, or of the excess of silver to gold in the preceding period, from 1781-1830. It must be evident to any impartial person that "the excessive output of silver" theory will not in any way account for the change in the relative value of the two metals.

Let us look at the figures relating to the output of gold and silver in another way (see *Colloquy on Currency*, by Lord Aldenham, p. 60).

Output of Gold 1850 to 1878 .	£662,000,000	
Output of Silver 1850 to 1878, valued at 15½ to 1 of Gold	312,400,000	
Excess of Gold (29 years)		£349,600,000
Output of Gold 1879-1892	£310,472,000	
Output of Silver 1879-1892, valued at 15½ to 1 of Gold	369,892,000	
Excess of Silver (14 years)		59,420,000
Excess of Gold (43 years)		£290,180,000

If we take the output of gold and silver from the beginning of the century, viz. from 1801 to 1894, we find that the gold produced has amounted to 3.015 tons, and the silver produced to 50.537 tons. This gives a ratio of 1 ton of gold to $16\frac{1}{4}$ (16.27) tons of silver. At this relative output silver should be at the price of 4s. 10d. the ounce, whereas the present price is about 2s. $7\frac{1}{2}$d. the ounce. At 2s. $7\frac{1}{2}$d. the ounce the relative output of silver to gold for the century at least should have been at the rate of 30 ozs. of silver to 1 oz. of gold, whereas we have seen that it has amounted to $16\frac{1}{4}$ ozs. only. Whatever way we take the figures, the output of silver cannot solve the question of the change in the relative value of the two metals.

The "output of silver" fallacy first gained currency in 1875, when the Comstock Silver Mines began to show great returns. It was then foretold that these mines would soon flood the world with silver, but before a few years had passed these mines began to produce gold, and in the result have produced silver and gold to almost equal amounts in value. The production of silver after gradually declining has now largely fallen off. After the Comstock Mines came the Great Broken Hill Mine of Australia to frighten those who were doubtful of a joint standard. This mine was spoken of as a veritable silver mountain where the precious metal could be shovelled out. The latest report from the mine shows that the estimated future output from this mine is something under £6,000,000. If it were possible to put it all on the market at once, instead of, as must be the case, over many years, it would not add more than about 7s. 6d. per cent to the world's existing stock of silver.

The truth is monometallists shut their eyes to the fact that infinitely greater variations in the supplies of the two metals occurred in the bi-metallic period prior to 1873 without disturbing the relative value of the two metals.

VII

"THE FALL IN PRICES IS ACCOUNTED FOR BY THE SUPPLY OF COMMODITIES OUTRUNNING THE DEMAND OWING TO INCREASED PRODUCTION AND BETTER FACILITIES OF TRANSPORT"

MANY monometallists lay stress upon the argument that the fall in prices is entirely a question of transport and distribution. "The rapid development of production and increased economy and facilities of distribution mean much more than the shibboleth of the bimetallists," writes Lord Playfair. This is one of those fallacies which contains an element of truth.

It is recognised that railways and other means of transport have increased. This is *that which is seen*. *It is not seen* that the increase in the facilities of transport has not caused any abnormal growth in the trade of the United Kingdom. The following table shows the increase in the railway mileage of the chief producing countries of the world:—

[TABLE.

TABLE III

RAILWAY MILEAGE (OPEN)

	1870.	1880.	1894.
Argentine	590	1,530	8,156
Australia	1,170	4,350	10,947
Brazil	630	2,100	7,492
Canada	4,010	6,145	15,627
India	4,775	9,308	18,855
Russia	7,100	14,700	22,590
Turkey	454	1,870	2,148
United States	44,614	86,497	179,279
Total	63,343	126,500	265,094

From the above table it will be observed that in *ten* years (1870-80) the railway mileage of the chief producing countries of the world doubled, and that the mileage doubled again in the ensuing *fourteen* years (1880-94). From Sauerbeck's tables we learn that the average price of 45 staple commodities fell from 96 in 1870 to 88 in 1880, a fall of 8⅓ per cent, but that between 1880 and 1894 the fall in the average price of the same commodities was from 88 to 63, or 28⅓ per cent. It might be possible, though unlikely, for the new railways built between 1880 and 1894 to have been instrumental in causing produce to be brought to our shores in amounts out of all proportion to our previous rate of trade expansion. But if we look at the following table we shall see this was not the case. On the contrary, the increase in the amount of our foreign trade, instead of showing a marked increase after 1873—as would have happened if "increased production and facilities of transport" had been the cause of the fall in prices—exhibits a loss of expansion. From 1860 to 1875 the tonnage of goods entered and cleared in the ports of the United Kingdom increased from .85 to 1.41 per head of the population, an increase of 76 per cent. But during the ensuing fifteen years from 1875 to 1890, the tonnage of goods increases from 1.41 to 1.98 per head, an increase of only 40 per cent.

TABLE IV

GOODS TONNAGE (SEA-BORNE) AND VALUE PER HEAD FOR THE UNITED KINGDOM

	Tonnage.	Value.				Tonnage.	Value.		
1860	.85	£13	0	0	1880	1.69	£20	3	0
1865	.90	16	8	0	1885	1.76	17	14	0
1870	1.17	17	10	0	1890	1.98	20	0	0
1873	1.39	21	4	0	1895	2.1	18	2	0
1875	1.41	20	0	0					

If, as monometallists say, the increased production of commodities, better machinery and greater facilities of transport explain the fall in prices since 1873, it is difficult to understand why this effect should only be first perceived after 1873, and why, when the world was endeavouring with equal energy prior to that period to put commodities upon the market as cheaply as possible, prices nevertheless rose. The "greater facilities" theory has one capital defect. It applies quite as much to the production of gold as it does to the production of other commodities. Since 1873 there have been many improved facilities for mining gold, fresh mines have been opened up and better methods of extracting the ore employed, yet the value of gold in purchasing power has been continuously rising.

The following table, compiled by Sir Guilford Molesworth, shows how, prior to 1873, prices rose with the increased output of commodities, and how after that date, although the previous rate of expansion in the production of commodities was not maintained, prices nevertheless fell away.

TABLE V

GENERAL COMMODITIES,
THEIR INCREASE OF PRODUCTION AND PRICES

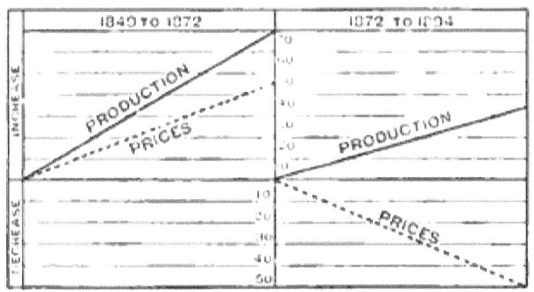

What magic was there in or about the year 1873 —other than the demonetisation of silver—to account for the striking change in general prices and the pertinacious rise in the purchasing power of gold which has gone on since that date. If the theory of increased production and greater accessibility of markets could account for it, the volume of trade would have shown at least as much expansive power after 1873 as it did before 1873, but instead of doing so, it exhibits a distinctly slower rate of growth.

Take again the following table relating to the railways of the United Kingdom :—

TABLE VI

BRITISH RAILWAY EARNINGS

	Miles.	Gross Earnings.	Average Earnings per Mile.
1853	7,569	£18,000,000	£2400
1873	16,083	55,000,000	3435
1893	20,646	76,000,000	3735

This table affords corroborative evidence of the check to the expansion of trade after 1873. In the twenty years preceding that date the net earnings of the railways of the United Kingdom increased annually at an average rate of £51 : 15s. per mile. In the twenty years succeeding 1873, the net average annual earnings increased at the rate of only £15 per mile. Railways are not generally credited with lowering their rates if the traffic will bear them, so that the diminished growth of mileage profits in the second twenty-year period is indicative of a less remunerative condition of trade.

Below are given returns of the total imports and exports for twenty years before and twenty years after 1873.

TABLE VII

BRITISH IMPORTS AND EXPORTS

	Total of Imports and Exports.
1853	£328,000,000
1873	682,000,000
1893	681,000,000

In these tables we have varied and corroborative evidence of the extraordinary check to the growth of English trade and of the break in prices since 1873. If these changes could really be accounted for by increased production and improved facilities of transport, we should certainly have seen the volume of trade as measured in tonnage (Table IV.) increase much more rapidly after 1873 than it did before that date.

The fallacy that the change in prices since 1873 is due to increased production was analysed in 1888 by Sir Robert Giffen (*Recent Changes in Prices*), and he gave it as his deliberate opinion that "about 1873 there was an alteration (in prices), but according to the best observation the movement in commodities continued what it had been, the quantity increasing at as great a rate as in the period just before, but not at a greater rate. The inference seems conclusive, therefore, that after 1873 the alteration in the economic movement was *in money*, and to this must be ascribed the change in prices which has occurred" (*Jour. Roy. Stat. Soc.*, 1888, p. 751). Similar evidence is borne by Mr. Mongredien, a most accurate observer, in his *Free Trade and*

English Commerce (Cobden Club), (1879): "We have adverted to the influence exercised over prices by the decreased production of gold and the increased volume of commodities which gold has to represent. The combined operation of these two agencies, while it has no effect upon exchangeable value, has produced, is producing, and will continue to produce a reduction more or less rapid in the money value of commodities."

Most monometallists who argue that prices have fallen, owing to increased facilities of production and transport and not owing to a restricted currency, argue in the same breath that the remonetisation of silver would "at once increase the price of the necessaries of life." As it is impossible for both these arguments to be correct, being mutually destructive the one of the other, we may feel additionally assured that the theory of increased production and transport cannot account for the stagnation which in recent years has characterised commercial enterprise in England.

VIII

"BIMETALLISM IS A QUACK REMEDY FOR BAD TIMES, AND BAD TIMES MUST COME NOW AND AGAIN"

THE fallacy that bimetallism is merely a quack remedy for bad times is one of those incomplete truths which readily deceive the unwary. Bad times will come from time to time, and no doubt the depression in trade which has prevailed in recent years has brought the question of a joint standard — whereby the world as a whole may recover a common currency — prominently before men's minds. But those who regard bimetallism from a scientific point of view, though they have used and rightly used, the evils which have resulted to trade from the loss of a common currency, to give point to their arguments, do not plead for a joint standard on that ground, but on the wide and impregnable basis of the need of a common currency for the whole world. An exactly parallel case is to be found in the Anti-Corn Law days. The potato famine in Ireland gave a tremendous impetus to the Free Trade movement, but no scientific Free Trader would base his argument for Free Trade upon a failure of crops, though he might fairly use such failure in illustration of the evils of protection. It would have been just as reasonable in the days of protective duties on food to use as an argument against Free Trade, that bad times must come now and again, as it is to use such an argument now against the remonetisation of silver. Such an argument wholly ignores the scientific aspect both of the Free Trade principle and of Bimetallism.

Taking the argument of the gold monometallists for what it is worth, let us ask them, when were times equally bad, and we are referred by a leading monometallist to the period from 1815 to 1845, "a period in which all the evils of which you complain existed in a more aggravated form than any of which we have now experience." It is somewhat curious that gold monometallists should pick out as a specimen period of bad times, the thirty years immediately following the demonetisation of silver by England, as the most nearly comparable with the thirty years immediately following the loss by the world of the joint standard. The admission that the period from 1815 to 1845 was a time of depression is important, for this period was essentially a period of contraction in the currency. Partly from a falling off in the yield of the mines, partly owing to the reaction from the forced currency of the preceding period, there was a contraction in the currency (1815 to 1845), and contraction is always mischievous. It looks as if, in spite of themselves, they were compelled to give evidence in favour of their opponents. But take the statement and compare it with Mr. Sauerbeck's figures. His figures show (they begin in 1820) that from 1820 to 1845 the fall in prices was at the rate of .9 per annum, but that from 1874 to 1894 the fall in prices was at the rate of 1.9 per annum, or more than twice as great. (See Appendix.)

It is not a question of bad times and good times, though these matters may fairly be discussed and given due weight. It is a question first of a stable, and secondly of a common currency for the whole world, which has been lost by the abandonment of the joint standard, first by England and later by the Latin Union. We are for a universal currency, and we accept a joint standard because a common currency upon a single metal is unattainable. We are for a universal currency based upon the joint standard, but those who call themselves monometallists would compel the one-third of the world to carry on its trade with the other two-thirds of the world by barter.

Trade considerably improved in the year 1890, and that improvement corresponded with the rise in the gold value of silver, to which we shall refer later (XVIII.), silver rising to 4s. 6½d. per ounce. An improvement in trade has again corresponded with a rise in the value of silver since the spring of 1895. Silver was then at 2s. 3d. per ounce. It has since risen to over 2s. 7d. an ounce—showing a rise of 13 per cent—and with this rise in silver a slight improvement in trade has followed. Although the purpose of bimetallism is to restore a common currency to the trade of the world, there is no objection to pointing out that the improvements observable in trade from time to time have been concurrent with a rise in the value of silver.

IX

"A COMMON CURRENCY FOR THE WORLD IS NOT IMPORTANT, FOR ALL TRADE IS BARTER"

WHEN Adam Smith elaborated the principle that "all trade is barter," he also made it clear that "barter was not sufficient to carry on the intercourse of mankind" and exemplified it at length. That "all trade is barter" is true in theory but not in practice. Barter is the exchange of commodity for commodity, and our experience teaches us that merchants who trade abroad do not carry on their business by exchanging commodity for commodity any more than we do in our daily life. That "all trade is barter" is true when we eliminate from the transaction the exchange of one article for money and the re-exchange of the money for another article, but the money as a measure of value "has," says Adam Smith, "become in all civilised nations the universal instrument of commerce." As between silver-using and gold-using countries there has been since 1873 no common money. The English merchant who sells his goods in India has in most cases to await payment, at least until he delivers the goods ordered and probably longer, and does not know what the value of the silver he has bargained to receive for his merchandise will, when he gets it, exchange for in gold. For as silver and gold are no longer embraced under one joint standard, he must either directly, or indirectly, through his banker, exchange the one metal for the other, as metal and not as money, and this is barter pure and simple. Owing to the loss of the joint standard in these latter days, money no longer exists as a universal instrument of commerce between "all civilised nations."

If any one is of opinion that the world needs no common currency because in theory "all trade is barter," let him try to imagine what ordinary life would be without money. Mr. Harvey has pictured it in the following words.

"If we had no money, one kind of property would be exchanged for another. Needing the calico on the merchant's shelf, you would exchange for it a bushel of potatoes or such property as you might have to offer. A sort of exchange value would be placed on all your property. A bushel of wheat would buy about so many pounds of sugar, and so on.

"Each merchant would have to be prepared to store all kinds of property, perishable and otherwise, he received in exchange for his goods. Railroads would have to arrange to receive payment for fares and freight in property, and store it until it again could be exchanged. If you went to the theatre you would have to take with you a crate of cabbage or some kind of property to pay your way in.

"There would be no practical method for paying labour. Commerce would virtually cease, and civilisation would go backward."

X

"THE UNITED STATES ARE A STANDING WARNING AGAINST BIMETALLISM"

THE opinion that the experience of the United States tells against bimetallism is a fallacy that has been widely spread by monometallists. As a matter of fact the experience of America tells in exactly the other direction, and proves up to the hilt the soundness of the bimetallic principle, viz. that a fixed ratio is capable of keeping the market value of gold and silver steady at that ratio.

The history of the currency of the United States divides naturally into two periods—the first bimetallic from 1792 to 1873, and the second monometallic since 1873.

The first of these two periods, from 1792 to 1873, divides again into two periods, viz. from 1792 to 1834, when the ratio was fixed at 15 to 1, and from 1834 to 1873, when the ratio was 16 to 1. Both of these ratios, it will be seen, varied from the ratio of the Latin Union, which, as we know, was $15\frac{1}{2}$ to 1, and the effect of this variation from the European ratio is most instructive.

Without troubling ourselves about minute fractions, we may say that at 15½ to 1 silver is worth 5s. 1d. an ounce, and that at 16 to 1 it is worth 4s. 11d., and at 15 to 1 it is worth 5s. 3d.

The difference between the ratio fixed in America and the ratio fixed by France operated in precisely the same way as the parallel variation between the English ratio and the French ratio in the last century, which we have already explained (II.). During the period from 1792 to 1834, when an ounce of silver was worth in the United States 5s. 3d. in gold and 5s. 1d. in Europe, all who had to make payments in bullion or specie between America and Europe by preference sent in payment gold to Europe and silver to America, as in Europe gold was the dearer metal, and in America silver was the dearer metal. Putting it another way, suppose that a Paris banker had to send specie to the States to balance accounts. At this time 15 ozs. of silver exchanged for 1 oz. of gold in the States, but in Paris it took 15½ ozs. of silver to exchange for 1 oz. of gold. The banker would naturally make his payment to America in silver. He would not ship gold to the States, for there each ounce of gold would exchange in America against 15 ozs. of silver only, whereas in Paris he could get 15½ ozs. for each ounce of gold. So if he had not a sufficient stock of silver on hand he would exchange gold for silver and ship the silver.

In the same way an American banker would not at that time ship silver to Europe when it required 15½ ozs. of silver to procure 1 oz. of gold. He would exchange his silver into gold in America, where for 15 ozs. of silver only he could obtain 1 oz. of gold.

The working of the American ratio took gold out of the country and brought silver in, simply because the difference in the ratio valued silver higher in the United States than it was valued in Europe. This proving inconvenient, the United States Treasury altered the ratio in 1834. The experience of the past forty-two years had not taught them the unwisdom of maintaining a ratio different from that of Europe, and instead of readjusting their ratio of 15 to 1 to $15\frac{1}{2}$ to 1, they erroneously changed it to 16 to 1, thereby undervaluing silver as much as they had overvalued it previously. The result was again precisely similar to their own previous experience and the experience of England in the last century, silver immediately began to flow to Europe and gold began to flow into the States, for the Americans gained 3 per cent by coining their silver into francs instead of into dollars. No doubt the United States had as an excuse for overvaluing gold, that they wanted to attract it back into the currency. "The bill was popularly known as the Gold Bill. Its advocates were jubilant and aggressive, and the purpose of over-valuing gold was announced." But it was unwise for all that, and the result was that gold rapidly returned to the States and silver as rapidly left. The want of small silver change became so serious that in 1853 Congress enacted that all fractional silver coins under a dollar should be made of light weight to prevent them from being exported. This was an entirely right principle to adopt, but at the same time the ratio should have been adjusted to the European ratio of $15\frac{1}{2}$ to 1. This would have assured the presence of both metals in exactly such proportion as they were required, for with international bimetallism if one metal is in greater demand for general convenience than the other, the moment the need of it is sufficient to pay carriage the metal needed comes. It is impossible with international bimetallism for either metal to vary in value from the fixed ratio more than enough to pay for its transit to the place where it is in demand. It is just as when over the bank counter the customer is asked "How will you have it?" and takes his money in silver or gold according to his requirements. So, with the re-establishment of the joint standard, each country would obtain exactly the proportion of either silver or gold that it required for currency purposes immediately the demand for either would cover the cost of transit.

The double experience of America, first one way and then the other way, following upon the experience of England in the last century (II.), shows admirably how promptly the two metals respond to a small variation in the ratio, and how, while such a variation in the ratio exists, the two metals will not circulate concurrently in the same country. We thus see the two metals obeying the law of the ratio with mathematical accuracy, absolutely unaffected by the variations in output, which, as Table II. (VI.) has shown, fluctuated immensely. Yet monometallists ask us to believe that the supply— that is to say the output—governs the relative value of the two metals, whereas, on the contrary, experience shows that the ratio, while it existed, governed with unfailing precision the relative value of the two metals.

The over-valuing of gold in the States from 1834

onwards, by fixing the ratio at 16 to 1, changed the States in effect from a silver-using into a gold-using country. Accordingly when, in 1873, France closed her mint to silver, it seemed to the States—just as in the case of England in 1816—an unimportant matter whether they closed their mint to silver or not, for silver had long ceased, owing to its being more valuable in Europe, to go to the United States mint for coinage. So when, in 1873, the United States closed its mint to silver under cover of a Codifying Act, which passed the Legislature without observation or remark, no one commented upon it, as to all appearance it was calculated to effect no change.

The experience of the United States shows, in a marked degree, how the slightest divergence from a common fixed international ratio is to be avoided, and how rightly John Locke and Sir Isaac Newton reasoned, when they decided that the true course to pursue, was to make the ratio identical with that of foreign states. Adam Smith also (*Wealth of Nations*, bk. 1, c. 5) pointed out the inconvenience arising from rating the value of silver lower in England than on the Continent : " In French coin and in Dutch coin 1 oz. of fine gold exchanges for about 14 ozs. of fine silver. In the English coin it exchanges for about 15 ozs., that is for more silver than it is worth, according to the common estimation of Europe." To remedy the inconvenience he advised " some alteration in the present proportion," in other words, in the ratio. The history of the United States currency from 1792 to 1873, instead of being a warning against bimetallism, confirms in a remarkable manner the truth of the international bimetallic principle.

The Gold Standard Defence Association assert that the experience of the United States "shows conclusively that no ratios between silver and gold have been able to maintain a parity of value." As a matter of fact, the experience of the United States exactly proves the contrary, and, as we have seen, shows :—

Firstly, that when there are two different ratios concurrently in force in different countries, each metal tends to go to the country where it is, relatively to the other metal, more highly valued. This was the experience of England in the last century. Gold came to England, where it was valued at 1 to $15\frac{1}{2}$ of silver, and silver went to France, where it was valued at $15\frac{1}{2}$ to 1 of gold. In the same way silver went to the United States mint prior to 1834, when the United States mint coined it at the ratio of 15 to 1; and gold went to Europe when it was coined by the Latin Union at 1 to $15\frac{1}{2}$ of silver. After 1834 gold went to the United States mint, where it was coined at 1 to 16 of silver; and silver went to Europe, where it was still coined at $15\frac{1}{2}$ to 1 of gold. Had the United States in 1834 fixed their ratio at $15\frac{1}{2}$ to 1, both metals would have come to their mint indifferently and the country would not have been denuded of silver after 1834, as it was of gold before that date.

Secondly, we learn, that so far from "the ratios being unable to maintain a parity of value between silver and gold," as the monometallists allege, the market value of silver and gold was absolutely unable to get outside the two ratios whatever might be the output. So long as the fixed ratios of the Latin Union and of the States continued, silver never

went above 5s. 3d. an ounce, and could not get below
4s. 11d. an ounce. While the ratio in the States
was 15 to 1 (before 1834) silver hovered between
5s. 1d. and 5s. 3d., and after the States altered their
ratio to 16 to 1, silver hovered between 4s. 11d. and
5s. 1d.—the price of 5s. 1d. per ounce in each case
being the value per ounce fixed by the French mint,
and 5s. 3d. the value of silver assessed by the States
prior to 1834, and 4s. 11d. the value fixed by the
United States subsequent to 1834.

Within these narrow limits of variation—these
limits of variation existing simply because of the
variation in the American and European ratios—
silver all these many years maintained its value
without a break, and followed the two ratios as truly
as the mariner's needle follows the pole.

This variation of the value of silver between the
two ratios—in the one case varying from 5s. 1d. to
5s. 3d., and in the other from 4s. 11d. to 5s. 1d., proves
beyond a doubt to any impartial mind, the absolute
character of the control exercised by the fixed ratios
over the market value of the metals. Yet the monometallists, pointing to these variations in value of silver
between the ratio points, assert that this variation
"proves conclusively that no ratios between silver and
gold have been able to maintain a parity of value," the
fact being that as between two different ratios such
a variation is inevitable, and that such a variation
only proves, that when countries establish a fixed
ratio between gold and silver, that ratio should be
identical. To prove the monometallist contention it
is necessary to show that silver went above or below
the limits of the ratios of France and the United
States. It is not sufficient to show that silver varied

in price within those limits. For variation within those limits only confirms the truth of the bimetallic principle. Yet this superficial fallacy has to do duty against bimetallism, and is passed current as an economic truth.

Since the closing of the United States mint to silver in 1873, the history of that country throws no light upon bimetallism, although gold monometallists often speak and write as if it pointed a warning against bimetallism. What are known as the Bland (1878) and Sherman (1890) Acts, did nothing for bimetallism. The Bland Act was intended to re-establish the fixed ratio, but when this measure reached the Senate, that chamber reconstructed the Bill and converted it from a bimetallic measure into an Act requiring the State to coin a given amount of silver every year. By this Act the Treasury was to coin not less than two million (£400,000) nor more than four million silver dollars annually. By the Sherman Act (1890) the Treasury was required to purchase 4,500,000 ozs. of silver annually. This imprudent enactment was repealed in 1895. It may be argued from the Bland Act that the action of upper chambers is not always wise. But neither the Bland Act nor the Sherman Act can be cited against the principles of bimetallism, for, on the one hand, neither of these Acts embodied in any way the leading bimetallic principle of a free mint with a fixed ratio, and, on the other hand, these Acts did embody the principle of fixed purchases of silver at stated periods by the Government, a principle which is entirely foreign to bimetallism.

The present position of the currency question in the States is one of an almost universal opinion in favour of bimetallism. Those who are in favour of gold monometallism are so few in number, it is confidently said, there would not be found a corporal's guard in Congress against bimetallism. The cleavage of opinion is not between gold monometallists and bimetallists, but between national bimetallists and international bimetallists. The international bimetallists correspond in the views they hold to English bimetallists, who would re-establish bimetallism upon an international basis, such as existed in the days of the Latin Union. The national bimetallists of the States are ready to re-establish it single-handed without waiting for the concurrence of any other nations. England, France, Germany, and the United States could undoubtedly re-establish the joint standard on as sound a basis as it existed before the break in 1873. In all probability France, Germany, and the United States could maintain a fixed ratio equally well, provided the Indian mints were reopened, and possibly this may be the ultimate solution. But anyway, those who represent the United States as divided in opinion upon bimetallism, and ask us to believe that the Republican party is averse to international bimetallism, utterly pervert the true facts. What the international bimetallists of the United States say is, that it would be quixotic on their part to throw themselves single-handed into the breach, when other nations are equally interested in restoring a common currency to the world.

When the English Press first directed public attention to the contest between Mr. M'Kinley and Mr. Bryan in the Presidential Campaign (1896), most of the newspapers spoke in terms of approval of Mr. M'Kinley as representing the "sound money party," oblivious to the fact that they condemned, generally in unmeasured language, English bimetallists who held similar opinions. Yet, judging from the usual comments of the English Press upon bimetallism, it is equally objectionable whether it comes in a national or in an international form.

It indicates some haziness in regard to bimetallism when newspapers, which comment with asperity upon international bimetallism as advocated in England, speak of the advocates of identical principles in the States as the supporters of "honest money."

The declaration of the M'Kinley platform was in these words :—" We are, therefore, opposed to the free and unlimited coinage of silver, except by international agreement, between the leading commercial nations of the world, *which we pledge ourselves to promote ;* and until such agreement can be obtained, the existing gold standard must be preserved."

Many English monometallists have spoken of the Democratic programme as if the payment of gold contracts in silver were a part of it. But this is inaccurate, for in his speech at Chicago which preceded his nomination, Mr. Bryan said, " Let me remind you that there is no intention of affecting those contracts which, according to present laws, are made payable in gold."

XI

"THE GRESHAM LAW (SO-CALLED) PROVES THE IMPRACTICABILITY OF BIMETALLISM"

ONE of the fallacies which some gold monometallists make use of for the purpose of "opposing bimetallism," is a curious application of what is known as the Gresham Law. Sir Thomas Gresham, in a letter written to Queen Elizabeth, explained the prejudicial effect to the State of debasing the standard coin, and stated quite truly, "that when persons are allowed to pay their debts in kind, they will naturally pay with the worst of their produce and keep the best, and if they are allowed to pay their debts in coin of different metals, the value of which differs in the market, but the value of which is rated equally in law, they will pay in the least valuable metal and keep the more valuable metal." From this it is argued with curious infelicity, that the coinage of silver and gold in the United States, first at 15 to 1, which drove out gold, and afterwards at 16 to 1, which drove out silver, was an example of the Gresham Law and "proved that the statute fixing the relative value of the two metals at these respective prices had no effect in restraining the operation of the Gresham

Law." The fact, as we have seen (X.), is, that the reason and the only reason why gold left the United States in the first period (1792-1834), and silver left the United States in the second period (1834-1873), was the existence of a second ratio in Europe, and between, and only between, these two ratios, could the price of silver oscillate. The Gresham Law could only operate over this narrow margin of $3\frac{1}{2}$ per cent between the two ratios. If it had any power to operate to a greater extent it is impossible to suppose that during a period of thirty-one years the variation in value between the two metals would not have been greater than $3\frac{1}{2}$ per cent, when within thirty years after the abandonment of the joint standard by the Latin Union and the United States, the value of silver measured in gold has changed from $15\frac{1}{2}$ to 1 to as much as 35 to 1, which means a variation of as much as 125 per cent. The wonderful stability of the market value of silver about the ratios—as long in fact as the ratios existed—proves, as we have shown (IV.), that the precious metals, when converted into currency and subjected to currency laws, differ from ordinary commodities, and that the enormous demand for these metals as currency, controls the market value of the uncoined metals.

XII

"BIMETALLISM SPELLS DISHONEST MONEY"

BIMETALLISM "is above all things a barefaced attempt to rob creditors for the benefit of debtors, and as such it is not surprising that it meets with unqualified opposition from all persons who care for common honesty." "It would not be the less robbery if it were international." So writes Lord Farrer (1896). Similar statements abound, but one will suffice. Certainly monometallists are emphatic.

We have travelled sufficiently far in our inquiry into the fallacies commonly current amongst gold monometallists, to be undisturbed by charges of "robbery" and "dishonesty." If, as Lord Farrer says, bimetallism is robbery and dishonesty, then "robbery" and "dishonesty" were rampant in England down to 1816, and were rampant elsewhere in the world down to 1873. What can have happened to make a return to the joint standard "robbery" and "dishonesty," unless the world's currency was tainted with "robbery" and "dishonesty" before 1873? We have seen that the operation of the joint standard down to 1873 inflicted no hardship on any one, on the contrary, its action was in the highest degree beneficial, for it not only provided the world with a common currency of which it stands sadly in need to-day, but it kept the value of gold steady relatively to the value of all other commodities. If hard words are appropriate to a discussion of this character, what is to be said of the morality which approves of doubling the indebtedness of debtors? Are creditors alone of any account in the world? Is it nothing that national, municipal, and private debts, contracted when gold had far less than its present purchasing power, must now be liquidated by a far larger outlay of human effort than either creditors or debtors ever contemplated? Indeed, so far as the national debt of Great Britain is concerned, it was wholly contracted before 1816, that is to say when the national currency of this country was on a bimetallic basis. The creditor is entitled to his pound of flesh, but the change in the value of gold gives him a claim to two pounds, and Lord Farrer says that, as bimetallism might have the effect of limiting him to his one pound of flesh, it is "robbery" and "dishonesty."

We have seen how, under the bimetallic system, the world possessed a common currency down to 1873, and we have seen how the loss of that common currency was the result of causes which may fairly be called accidental. We have seen further how the loss of that common currency has been followed by a persistent breakdown in values, and we have seen also that, examine the matter in what way we will, there is nothing in the output of silver or in any extra increase in the output of commodities, or in the means of transporting them to market, to account for the fall in prices which has been brought about. We have seen how the change in the relative value of silver and gold "has laid the white labourer, paid in yellow money, open to more severe competition with the yellow man paid in the white metal than in the past." When we propose, taking all these considerations into account, to go back to the system which obtained with complete success prior to 1873, we are told that it is a "barefaced attempt" at "robbery" and "dishonesty."

The abiding cause of the great change in values and the diminished growth of trade and commerce is the restriction of the currency by the abandonment of silver. The demonetisation of silver alone can account for the changes which have come over the trade and commerce of the world since 1873. With a growing population and a growing trade, silver was suddenly abandoned by Germany, France, Italy, Belgium, Switzerland, and the United States, while simultaneously Russia and Austria bid for an increased store of gold. This vast and violent restriction of the currency has naturally greatly enhanced the purchasing power of gold. But to attempt to return to the old and well-tried paths of a joint-standard of currency is stigmatised as "dishonest."

The highest estimate of all the gold in use as money puts it at £880,000,000. That is to say, all the gold used for money occupies only 11,000 cubic feet of space, and can all be put into a room measuring 22 feet 6 inches in length and in breadth and in height. The whole of Western Europe, the United States, and the English Colonies are carrying on their vast commercial transactions upon a monetary basis which—when we have eliminated paper money

—consists of a single block of gold measuring less than 22½ feet each way. The wonder is, not that the restriction of the currency since 1873 has caused changes in values, but rather that those changes have not been greater.

Sometimes we hear men talking of the flood of silver. The highest estimate of the silver in use in the world as money (valuing silver at the old ratio of 15½ to 1) puts it at £870,000,000. If we put it as high as the estimate made of the stock of gold used for money, viz. £880,000,000, it will all go into a cube measuring 69 feet each way.

As Germany, France, the United States, and other countries abandoned silver, the demand for gold increased, and "as the standard for measuring the values of commodities in these countries is now only half what it had been, the decline in the value of commodities followed."

Another form which the fallacy of "dishonest money" takes is met with in the saying that bi-metallists "want to turn a shilling into eighteen-pence." The silver shilling used to be worth about a shilling. The crown piece (5s.) weighs an ounce, and before 1873, if hammered up and sold as silver, it would fetch almost five shillings. It was not quite worth it, being slightly alloyed to enable it better to stand wear and tear. The re-establishment of the joint standard at a ratio anywhere near the old ratio of 15½ to 1 would not change a shilling into eighteenpence, but would at most bring up the present shilling, which is worth barely sixpence, to an honest shilling, as being worth somewhere about the value it has stamped upon it. Bimetallists seek a common currency upon an honest basis. They do not want dishonest but honest money. What is the essential characteristic of "honest" money? Stability of value. And silver has shown greater stability of value than gold, as we have already seen (V.).

As it is strenuously asserted that silver is not "honest money" let us here note the four qualities which economists agree are essential to a sound and honest currency. In theory one substance is as good as another as a medium of currency, but in putting various substances to the test of the four canons of a sound currency, only two commodities meet the conditions laid down, namely, gold and silver. These conditions are—

> Easy divisibility,
> Intrinsic value,
> Non-deterioration by keeping, and
> Stability in value.

Bread is easily divisible and possesses intrinsic value, but it deteriorates by keeping. Jewels possess intrinsic value and do not deteriorate, but are not divisible. And so, in different ways, all other commodities, save gold and silver, are found wanting in some one of these essential characteristics. By tacit consent of all nations silver and gold have been fixed upon as alone fulfilling these conditions.

When we ask ourselves which of the two metals, silver or gold, best fulfils the conditions laid down, we find that as to the first three conditions they may be ranked equally, but as to the fourth condition— stability in value—silver, judged by any test we can apply, shows greater stability than gold. The purchasing power of silver has remained constant while the purchasing power of gold has increased. Silver, therefore, if we are to judge between the two metals and use the term honest, is the more honest money. It has neither defrauded the debtor by compelling him to return more than he has borrowed, nor has it given an unmerited bonus to the lender. Silver is, by all the tests and conditions of the economists, essentially "honest" money. But there is a better money than either gold alone or silver alone. It is furnished by the two metals joined in the duty of providing money as under the English law of the last century and the French law of the present century. Such a money is a better money because, being on a broader basis—just as a man stands more firmly on two legs than upon one—its variations in value are necessarily less than the variations in value of a monometallic money. As more stable in value it would conform more closely than either gold alone or silver alone to the requirements of a sound currency and therefore be a more "honest" money.

XIII

"AS THE BANKS AND TREASURIES OF THE WORLD ARE OVERFLOWING WITH GOLD, IT IS ABSURD TO ARGUE THAT GOLD IS SCARCE"

THE argument in favour of monometallism, founded upon the allegation that the banks and public treasuries of the world are overflowing with gold, is a specious one but unsound. It is another of those common fallacies, the truth of which is seen, but the untruth of which is not seen. *It is seen* that the metallic reserves of certain banks have largely increased. *It is not seen* that while banks have increased their metallic reserves, the aggregate deposits of customers have become almost stationary. We have just shown how the fall in prices and profits, without a corresponding fall in the rate of interest for loans, has checked borrowers from embarking upon fresh enterprises and new ventures—in the words of Mr. Balfour, "benumbed the springs of enterprise." The banks are overflowing, not with increased deposits, but with money called in because profits are now so slender, capital can no longer be loaned without risk. The rate of interest has fallen because of the widespread indisposition to enter upon new undertakings. The general commercial stagnation characteristic of recent years, having on the one hand caused banks to curtail their advances, and on the other hand daunted men from launching out in new undertakings. This explains the increase of unemployed balances, and instead of being a healthy is, on the contrary, an unhealthy symptom. The published returns of the aggregate deposits in banks have shown large *apparent* increases, owing to the conversion of many private banks into joint-stock companies, their deposits thereupon being incorporated in the published returns.

When the business of the country is really prosperous there is a steady accumulation of fresh capital, and a portion of this accumulation is added to the aggregate deposits in banks, which in former years used to show a steady increase year by year.

We have said that the fall in the rate of interest has not corresponded to the fall in values. This is sufficiently apparent from the price of consols. The fall in prices has enabled the Exchequer to convert consols to a lower rate of interest, from 3 per cent to $2\frac{3}{4}$ per cent in 1888, and to $2\frac{1}{2}$ per cent in 1903. That this relief to the nation as taxpayers, is not commensurate with the fall in prices from par to 68 per cent (V., Table I.), is apparent from the market price of consols. When the world enjoyed a common currency prior to 1873, and trade and commerce were increasing "by leaps and bounds," the average price of consols was about £92, as will be seen from the table on the following page. From 1860 to 1870, when trade was good, consols were very stationary, about £92. But as prices fell and commercial enterprise was checked, in

TABLE VIII

Price of Consols

1860-64 average price						£92¼
1865-69 ,, ,,						91½
1870 ,, ,,						92½
1875 ,, ,,						93⅞
1880 ,, ,,						98⅞
1885 ,, ,,						99½
1890 [1] ,, ,,						96¼
1895 ,, ,,						110½

[1] Interest reduced to 2¾ in 1888.

spite of a reduction of interest, consols rose 20 per cent in value. In some degree, no doubt, the rise in value is due to the large purchases on account of the Post Office Savings Bank, but here again the increase in deposits indicates an inability to employ capital with confidence. The ease, too, with which municipalities can now borrow money at 3 per cent, and in some cases even on more favourable terms, instead of being called upon to pay 4 or 4½ per cent, the usual rate a few years back, is another sign of want of confidence.

It is evident that the reduction of ½ per cent in the interest on consols has not compensated the taxpayers as debtors for the increase in the burden of the national debt.

Take again the rate of interest on money lent upon mortgage. Four to four-and-a-half per cent was the usual rate according to circumstances a few years ago. Since the fall in prices mortgagers have been able to renew their mortgages at from 3½ to 4 per cent. But here again the $\frac{1}{2}$ per cent, possibly $\frac{3}{4}$ per cent, which they have saved, has not adequately compensated them for the losses they have sustained from the fall in prices.

India is a debtor, and a large debtor to this country. India has to make home payments, as they are called—that is to say, payments to Great Britain on Government account—of about £20,000,000 annually. With the silver rupee exchanging at two shillings, India had to furnish 200,000,000 rupees annually to keep up those payments. With the rupee at half that value, she has to bear a double weight of taxation for this purpose, in order to furnish 400,000,000 rupees. But then India is our debtor, and debtors are of no account. (See also as to the gold supply, Sec. XXVI.)

XIV

"BIMETALLISM IS ONLY A NOSTRUM TO HELP DEBTORS"

THE fallacy that bimetallism is "only a nostrum to assist debtors," involves the idea, already considered, of a dishonest money (XII.). But the particular form this fallacy takes, aims more especially at "debtors," and is framed in such a way as to imply that debtors, as a class, are not desirable persons to assist. In short, it implies that a fall in prices, in so far as such a fall affects debtors, is a matter of no importance. But there are debtors and debtors. In one way we are all debtors. We all have to pay taxes to the State, and in many cases rates to local authorities, which are being applied to pay off public debts contracted before the abandonment of the fixed ratio. Money borrowed for public and private purposes is borrowed, as a rule, over many years, and the rate of interest very slowly responds to a fall in prices.

Let us, however, take the case of the ordinary debtor, and ask ourselves whether, if the joint standard be re-established, in order to provide the world once more with a common currency, we need be wholly unconcerned if, as a subordinate result of that change, the ordinary debtor feels some relief from his burdens. As we have said, there are debtors and debtors. The improvident debtor who wastes his substance upon his pleasures, is not a subject for special sympathy. But these are few in number compared to the vast class of worthy debtors, who, by their labour and their intelligence and their enterprise, are, with the aid of a certain portion of borrowed capital, creating and maintaining trade and business throughout the community. The balance-sheets of banks prove that some 50 per cent of the money deposited with them by customers, is lent out again to enterprising but provident borrowers. And the vast sums lent out by the banks, bear but a small proportion to the total amount of money lent out on credit, and employed in trade in a similar manner. Lenders who sit at ease, having no care but to receive their interest at stated intervals from the hard-working borrower, would fare badly, if borrowers were not to be met with to employ their capital. With the fall in prices, borrowers have found their burdens in the way of interest heavier than before; profits have run down, but interest charges have remained. The result is, that they are unwilling and unable to borrow at the old rates of interest (XIII.), and the rate of interest has gone down. Yet, in spite of the drop in the rate of interest, prices are in many businesses so low and profits so uncertain, that lenders are reluctant to embark upon new ventures.

XV

"LOW PRICES ARE NOT AN EVIL, ON THE CONTRARY AN ADVANTAGE"

THE "WEIGHTS AND MEASURES" FALLACY

"Low prices are not an evil, on the contrary they are a great advantage to the general mass of people." This is another proposition which seems at once to carry the ready assent of every one. The advantage of low prices for the necessaries of life is that *which is seen*. *It is not seen* that the general mass of the people are workers, and that, as workers, if prices are low, they will probably find part of their occupation gone. When the prices of general commodities remain steady, and the prices of a few particular commodities fall, the cause may almost certainly be attributed to improved facilities in manufacture and output. But when the prices of commodities in the mass show a general decline, which, as we have seen (V., Table I.) is characteristic of prices in recent years, then it can only advantage us, that prices have fallen, because the fall has been unequal, and has not reached or affected the particular occupation in which we are employed.

In theory it is unimportant whether the value

of commodities rises as a whole or falls as a whole, since the commodities still retain the same relative or exchangeable value for each other, whether that value is expressed in higher or lower monetary values. But as regards individuals and classes, the effect of a rise or fall in values is considerable, for in practice, a rise or fall in the purchasing power of money, as reflected in the lower or higher prices of general commodities, does not take immediate effect upon a very considerable number of incomes.

We have just referred (XIII.) to the slow readjustment of interest upon loans. The recipients of fixed incomes generally, have only in part experienced the full effect of the change of values. But the recipients of fixed incomes can hardly be regarded as "the general mass of the people"; they are rather the happy exceptions. The general mass of the people are wage-earners, and they are directly affected by lower prices. Lower prices are generally adverse to a revival of trade. Capital and credit, the two wheels on which commerce revolves, and without which it drags, keep aloof from falling markets, and distrust the security of property that is declining in money value. It is when prices are rising, that capital and credit come forward freely to accelerate the rise. Capital and credit associate with prosperity and improving trade. It is best, however, in the interests of all, that prices should remain as steady as possible, at any rate that they should not be liable to *monetary* changes of value, in addition to the fluctuations to which they are inherently liable. But when increasing trade requirements are met by a diminishing supply of the medium of circulation, from the closing of many mints to silver, the stability of prices must be seriously affected independently of all ordinary causes.

David Hume, in his *Essay on Money*, says, "If prices rise, everything takes a new face, labour and industry gain life, the merchant becomes more enterprising, the manufacturer more diligent and skilful, and even the farmer follows his plough with greater alacrity and attention. If prices fall, the poverty, beggary, and sloth that must ensue are easily foreseen." Professor Stanley Jevons, writing on the same subject in 1862, endorsed the opinion of M'Culloch in the following terms: "I cannot but agree with M'Culloch that, putting out of sight individual causes of hardship, a fall in the value of gold must have the most powerfully beneficial effect. It loosens the country, as nothing else could, from its old bonds of debt and habit. It throws increased rewards before all who are making and acquiring wealth. It excites the active and skilful classes of the community to new exertions, and is, to some extent, like a discharge from his debts to a debtor long struggling with his burdens."

Falling prices brought about by a rise in the value of the medium of circulation, can benefit no one unless he is able to resist a proportionate abatement of his income. The well-to-do classes, who derive their incomes mainly from investments, are best able to resist abatement of their incomes. Certainly "the mass of the people" are not able to do so long. They may temporise by working short time, but the inevitable fall in wages must come. In employments where the rise in value of the medium of circulation lays those employed open to competition with labour paid in a metal which has not risen in value—as in the silver countries—they will be fortunate if their occupation does not slip away from them altogether. In the words of Sir James Graham, "the annuitant and the tax-eater rejoice in the increased value of money, in the sacrifice of productive industry to unproductive wealth, in the victory of the drones over the bees."

Some monometallists, feeling pressed by the argument of "low prices," assert that "low prices are not a monetary question at all," and that "money is a measure, not a creator of prices." Money is certainly a measure of values, but it is a fallacy to suppose that "money cannot affect prices." The confusion arises from regarding money as a foot-rule or a pound weight. A foot-rule is of a fixed length and a pound weight of a fixed weight, both established by law and unchangeable. But when legislation creates money, it can only define its weight and fineness, and enact that it shall be accepted as legal tender; it cannot define what shall be its purchasing power. (See also Sec. XXII.) The neglect of this distinction —that the measuring value of money, unlike other measures, cannot be regulated—keeps alive the fallacy that "money cannot affect prices." So long as the metallic currency remains fairly constant in quantity, the effect of money upon prices may be disregarded. But when there is a distinct expansion of the currency, as in the gold-using countries in the 'fifties, or a distinct contraction of the currency, as in the same countries after 1873, when the world's currency was snapt asunder, then prices become subject to direct monetary influences.

Mill wrote: "As the whole of the goods in the

market compose the demand for money, so the whole of the money constitutes the demand for goods." In 1873 many countries deprived themselves of silver as legal money ; accordingly since that date " money constituting the demand for goods " has diminished in gold-using countries, and goods have fallen in price. The diminished volume of metallic money available as legal tender, has accordingly caused money in gold-using countries to become an active " creator of fresh prices," and, as the legal monetary tender has diminished, of necessity, a creator of lower prices.

"Industry and commerce," says Mr. Leonard Courtney, "can in the end adjust themselves to any scale, but consider what is involved in the process of adjustment. Suppose gold so enhanced that a quantity of goods that sold for £100 will now fetch only £50. The maker, out of his £100, paid national charges (including interest on debt), local charges (with similar inclusion), probably interest on borrowed capital, cost of raw materials, wages ; and something was left over for his own profit. The item in the cost of raw materials may be assumed to be reduced one half, but this shrinkage does not apply to the other items at all, or not with the same quickness. National debts, local debts, do not change ; salaries of functionaries change very slowly ; there is a good deal of friction before wages fall. The profit of the manufacturer is the residue which must suffer all the reductions that *cannot be passed on*. The readjustment involves a grinding and grating of the parts of the industrial organism, and a constant contracting of industry to escape or minimise loss. The fallacy lies in the assumption that the shrinkage due to an appreciation of gold is uniform and immediate all round. Wages do not fall so rapidly as prices, and therefore the wage-earning classes temporarily benefit ; but the gain is counterbalanced by lessening enterprise and activity, from which, in the end, labourers suffer. If an international agreement established a ratio, and thus ensured a universal standard of value, that standard would still be liable to change ; but the change would affect all nations, so as to remove the present aggravation of international competition ; and as, moreover, the standard of the future would be controlled by the metal which is tending to become cheaper, the effect of the changes of the future (common to all) would be to stimulate, rather than contract, enterprise—to make industry everywhere more energetic, because diminishing the weight of the charges upon it. For these reasons I believe the gain that would ensue for England, would far outweigh any loss of tribute on the part of money-lending Englishmen."

At the British Association (1895), Dr. Smart stated the result of low prices to the following effect :—Manufacturing differs from commerce in that it cannot be healthy or prosperous with falling prices, because all manufacturing has a history behind it. The manufacturer's capital is mostly fixed, and is subject to charges that do not vary with falling prices. Thus his position is always difficult, as improvements in process recur very frequently, and rapidly superannuate existing machinery. If beyond this continual risk something else depresses the prices of his goods, while the dead weight of fixed capital hangs round his neck, how is the manufacturer to prosper? Now, who are the manufacturers? Are they not the leaders of the national industry, the "employers" of those who labour? How can there be prosperity in national industry when the very heart is taken out of those who set it to work?

XVI

"BIMETALLISM WOULD DEPRECIATE THE CURRENCY, AND TO DEPRECIATE THE CURRENCY IS TO ROB LABOUR"

THE fallacy contained in the statement that gold monometallism is best for labour, lies in the word "labour." "Labour" used in this general sense, means labour as a whole—means that if a profit and loss account is struck, between the takings and outgoings of all who can be classed as labourers, a balance will come out on the right side. Because certain of the labouring classes have been able to retain their occupations at their old rate of wages, and have apparently profited by the fall in prices, we are invited to assent to the proposition, that labour as a whole has profited by the loss of the joint standard. It *is not seen* (1) that large numbers of labourers have been driven from their employment; (2) that a large number of others have had to submit to lower wages; and (3) that those who have apparently profited by the fall in prices, would, under the continuance of the joint standard, have been in a position to obtain a rise of wages.

It is an ill wind that blows nobody good, and here and there small classes of labourers may have actually profited by the loss of the joint standard. Labourers employed in agriculture may lose their employment, while the hands engaged in breweries, where profits have risen largely as agricultural produce has fallen in value, may not only have retained their old rate of wages, but may have had their wages raised.

It is possible for a time, but only for a time, for labour to profit by a fall in prices. This is due to the well-ascertained fact that wages, unlike commodities, do not readily respond to a rise or fall in prices. Merchants can mark up or mark down the price of their goods, without exception being taken, but when they attempt to lower wages the wage-earners resist any reduction. Employers of labour accordingly practise every possible economy before resorting to a reduction in wages. On the other hand, knowing how difficult it is to reduce wages once raised, they hesitate to raise them, until assured that higher prices or profits appear permanent.

With the rise in the price of commodities which took place between 1850 and 1873, the wages of labour rose also. Labour organised itself in trade societies and unions, and artisans and labourers shared in the general improvement. Every capable man wanting employment found a place. In a paper read by Sir Robert Giffen before the Statistical Society in 1886, he showed that the rise in money wages between 1836 and 1886 had been from 50 to 100 per cent, and that this improvement appeared to culminate about 1873. Improved machinery enabled the turn-out of our cotton, woollen, iron, and steel mills to be increased during that period, threefold and fourfold, yet prices and wages continued to rise. Since the fall in prices began after 1873, wages have fallen in almost every employment which is not more or less in the nature of a monopoly, such as railways, the liquor trade, the post office and police.

In the United Kingdom, since the fall in agricultural prices after 1873, 2,500,000 acres have been laid down to grass. This change, from arable to grass, has displaced the labour of at least 50,000 agricultural labourers, who have had, with their wives and families—in all about a quarter of a million persons —to betake themselves to other occupations or to other lands. In Cornwall the tin miners have found their industry slipping away, as mine after mine has been shut down. When silver was 5s. an ounce it was necessary to send £120 in gold to the Straits Settlements, to purchase a ton of tin. Then the Cornish miners thrived. Now £57 in gold will purchase as many silver dollars as £120 would formerly. Accordingly the Straits tin can be put upon the English market under £60 a ton, and the Cornish miners lose their employment. In the Cotton-spinning industry, we find our best customers in the East (India, China, and Japan) unable to purchase from us as before, being restrained by the lesser purchasing power of their silver currency. Their purchases are reduced, and they are importing cotton-spinning machinery to manufacture with their silver-paid labour the fabrics they need. A return recently submitted to the Manchester Chamber of Commerce, shows that the shares of only 6 out of 93 spinning companies were selling at a slight premium, while 52 of these companies were paying no dividends on their capital. At the same time the British Consul at Hiogo reports that the Japanese cotton mills are running at average profits of nearly 20 per cent.

Now if all these changes were the outcome of a real variation in the relative output of gold and silver, however much we might regret the resultant losses to labour, we should nevertheless feel that they were in the nature of the inevitable. But this is just what is not the case. The whole of these dislocations of values and of labour are the direct result of legislative action (IV.). There has been no change in the relative output of the two metals (VI.) such as can in any degree account for, or justify, the alteration in their present relative market value. What possible justification can gold monometallists give, not only for the loss of a common standard of currency, but for all these attendant evils resulting in widespread displacements of labour, reductions of wages, and loss of employment.

Monometallists assume that bimetallism would depreciate the currency, but it could only depreciate the currency if an inequitable ratio were determined upon, and this is a gratuitous assumption. Bimetallists have no desire to re-establish any ratio other than a just ratio, but of this we shall speak presently. It is sufficient here to refer to what has already been said (VI.) as to the proportionate output of the two metals in the present century.

Monometallists appear to consider, when they have set out a few figures showing that in the last twenty years there has been an increase in the average general consumption of common food, that they have proved conclusively the advantages of gold monometallism.

TABLE IX

(Average Consumption of Articles per head of Inhabitants of the United Kingdom in lbs. and 100ths of lbs.)

	1875.	1885.	1895.
Butter and margarine	...	7.21	10.44
Beer (gallons)	...	27.0	29.0
Cheese	5.46	5.53	6.40
Coffee	.98	.91	.68
Currants	4.29	4.16	4.90
Pork	...	1.14	1.13
Sugar	62.0	74.0	79.0
Tea	4.44	5.06	5.52
Tobacco	1.46	1.46	1.60
Wheat	197.0	189.0	201.50

This table does not show any marked increase. If we had the necessary statistics at hand to enable us to tabulate the average consumption of the same articles, say in the year 1850, it is probable that we should see indeed a striking change, and that the gold monometallists would not be very proud of the increase which is shown between 1875 and 1895. It is known, however, that the average consumption of sugar was 25 lbs. per head in 1850, so that the great increase in the consumption of sugar took place when the world had a joint standard. We know also that the consumption of tobacco amounted to 14 ozs. (.88 of lb.) per head in 1840, so that here again the great increase in the consumption took place when the mints were open to silver.

To re-establish the joint currency is to call back silver into monetary use as legal tender, and to remonetise silver is to enlarge the metallic currency. We have already referred to the beneficial influence of the increased output of gold following the Australian and Californian discoveries, and now quote the opinion of Mr. William Newmarch (*New Supplies of Gold*, 1853, pp. 89-90) on the effect upon labour of an enlargement of the metallic currency. " We are justified in describing the effects of the new gold as almost wholly beneficial. It *has* already *elevated the condition of the working and poorer classes;* it has quickened and extended trade, and exerted an influence which, thus far, is beneficial wherever it is felt."

Nothing shows more accurately the effect of restriction of the metallic currency, than the tables of Mr. Sauerbeck (see Appendix). These tables, as Professor Foxwell has said, are " the bed-rock of all profitable monetary discussion." They show us that prices measured in gold have fallen, and it must follow from this that gold has appreciated in value. But Lord Farrer, who is a leader amongst monometallists, has endeavoured to decry these tables by saying, " If there could be anything like an ideal standard of value it would be a fixed quantity of labour " (1894). This implies two things—that there is " no ideal standard of value," and that the best form of it would be " a fixed quantity of labour."

Professor Foxwell has pointed out in an admirable criticism (*National Review*, Jan. 1895) that, in the first place, so far from Sauerbeck's tables not being an ideal standard of value, the British

Association in 1877 appointed a committee of experts to consider and report upon index-numbers, and any one who reads Sir Robert Giffen's report (1889) will find, that the committee substantially confirmed Mr. Sauerbeck's method.

In the second place, labour could never be "an ideal standard of value." Labour, as Mill has observed, is not a measure of exchange value but of national progress. Jevons agrees, for he wrote, " I hold labour to be essentially variable, so that its value must be determined by the value of the produce, not the value of the produce by that of the labour." Sir Robert Giffen has also stated that in an historical comparison it is commodities which appear as the fixed unit, and labour as the variable unit.

Professor Foxwell has further shown that the notion that "labour" can be regarded as a commodity has arisen from want of thought. Labour does not become a commodity because it "happens to appear on the same side of an employer's balance-sheet as the coals which heat his boilers." Thus in one short sentence Lord Farrer lays down two absolutely erroneous propositions.

The assertion—baseless as we believe it to be— that the re-establishment of the joint standard would rob labour, is one of those side issues calculated to lead men into an interminable discussion upon profits and losses, forgetful of the main question, namely, the utility of and the practicability of re-establishing a stable and a common currency.

The leaders of labour at any rate do not appear to be enamoured of the doctrine of cheapness, and prefer constant employment and good wages as the more advantageous to the working classes generally, even if accompanied by a somewhat higher level of prices. They know "how unprofitable it is," to quote the words of Sir Josiah Child (1690), "for any nation to suffer Idleness to suck the breasts of Industry."

XVII

"SAVINGS BANK RETURNS, SCHOOL, CRIMINAL AND OTHER STATISTICS PROVE THAT ALL IS WELL"

WHEN a large body is moving with a considerable impetus, it does not, when the propelling force is withdrawn, lose its momentum all at once. It is not surprising, therefore, if in many ways it can be shown that steady progress is being made even in a time of commercial lassitude. Statistics however, as to Savings Bank returns, schools, prisons, and so on, raise at best but side issues whichever way they may appear to tell.

A broad general view of English trade and commerce shows us that in the quarter of a century which elapsed between 1849 and 1873, with free trade and —thanks to the Latin Union—the full benefit of mints open to both silver and gold, the trade and commerce of England made amazing strides, and that in spite of a Russian War and an Indian Mutiny, taxation was steadily lightened, while with a rapidly increasing production of commodities, prices rose (VII.). On the other hand, in the quarter of a century which has followed 1873, with no war rising above a frontier skirmish, with a slackened output of commodities, but with mints closed to silver, prices have fallen, trade has languished, and taxation has tended rather to increase than to diminish.

Satisfactory explanation of this strange state of things is to be found, according to gold monometallists, by studying certain statistics relating to bank deposits, schools, and prisons. But the difficulty we have in drawing any conclusions from such statistics is, that at the dates selected for comparison different conditions existed. The Post Office Savings Bank, for example, from time to time has relaxed the limits fixed for the amount any one person may deposit in one year. The result is that it is impossible to say what portion of the increase in the total amount of deposits may not be due to increased facilities for deposit. Moreover the growth of the Government Savings Bank may be considerable even in a time of depression, as affording a refuge for capital at a time when other securities are causing uneasiness.

Some have appealed to the growth of the Post Office returns and the great increase in the number of letters transmitted by post, but such statistics are valueless during a period when elementary education is multiplying letter-writers day by day.

The increase in the number of scholars attending elementary schools is the result of legislative and departmental activity. While the decrease in the number of criminal convictions is in part the result of a better educational system. In 1872 we spent £1 : 8s. annually upon the education of each child attending an elementary school, and at that time large numbers attended no school. Now we spend £2 : 3s. annually upon elementary scholars, and few can evade attendance. In addition to this influence, during the past twenty-five years the police system of this country has become more efficient year by year.

Some have had the courage to appeal to the income-tax statistics, and if the income-tax returns exhibited buoyancy it would be an argument that, however inconvenient to the world the loss of a stable and a common currency might be, no great evil had been suffered by the people of the United Kingdom. But the income-tax returns speak with no uncertain sound as to the change which has taken place.

TABLE X

INCOME-TAX RETURNS

Year.	Population.	Each penny in the £ produced	Each penny in £ equal per head of the population to
1853	27,000,000	£ 813,000	7¼d.
1873	32,000,000	1,875,000	1s. 2d.
1893	38,000,000	2,190,000	1s. 1¾d.

It is said that if due allowance be made for the abandonment of income tax on small incomes, the tax now would show a yield of £2,250,000 for each penny, equal to 1s. 2¼d. per head of the population. There are two observations which may be made as to this. Firstly, the rigour with which the income tax is now exacted is very different from the way in which it was collected in 1855 or even as lately as 1875. It is probable that its more punctual exaction has much more than compensated for the loss of the tax upon small incomes. Secondly, even if we allow this readjustment, it only brings the produce of each penny up to 1s. 2¼d., whereas if the improvement shown between 1853 and 1873 had been maintained between 1873 and 1893, each penny would have produced per head of the population 2s. 3d. in 1893.

XVIII

"WANTED A RATIO? BIMETALLISTS HAVE NO DEFINITE PLAN"

IN 1895 Lord Farrer, a leader amongst the gold monometallists, announced that "the bimetallists had no definite plan." In 1896 Lord Farrer abandoned this theory for another, which is that the bimetallists have such dishonest intentions that they dare not reveal them. Let Lord Farrer speak for himself. "The controversy as to 'wanted a ratio' made it clear that what our bimetallists really desire is that there shall be free coinage of silver at a ratio of 16 ozs. of silver to pay a debt of 1 oz. of gold. If they do not propose this in so many words they know that such a proposal would shock the conscience of the public" (25th July 1896).

Here we have two definite statements, first that English bimetallists desire to re-establish the ratio at 16 to 1, secondly that such a ratio would be a dishonest one.

Let us in the first place deal with the honesty of such a proposal, if it were made. We have already shown (VI.) what the relative output of silver has been for the last two centuries, and what has been the actual output of the two metals during the present century. Many monometallists seem to take it for granted that the relative value of the precious metals follows their relative output. They then picture to themselves a vast output of silver, and jump to the conclusion that the present price of silver (measured in gold) is the natural result of that output. But assuming for the moment that the relative value follows the relative output, what are the facts. During the present century the actual output of silver to gold has been $16\frac{1}{4}$ ozs. to 1 oz. of gold. At $16\frac{1}{4}$ ozs. of silver to 1 of gold, silver would be worth 4s. 10d. per ounce. If, however, the ratio were to be fixed at 16 to 1, silver would be worth 4s. 11d. per ounce. Accordingly if the ratio of 16 to 1 were adopted, silver would be over-valued to the amount of 1d. per ounce. The idea of such a thing being done shocks the moral feelings of Lord Farrer, but that silver should now be passing current at 2s. 7d. per ounce (equivalent to a ratio value of $30\frac{1}{2}$ to 1), not owing to any increase in the output that could by any possibility account for such a change in value, but simply due to legislative enactments, does not shock him at all. The debtor may be shorn of his last shred of wool, but the creditor, although by accident entitled to more than was ever contemplated, is to have his unearned increment preserved inviolate. It is courageous of Lord Farrer to appeal to moral principles when a creditor's penny is in danger, yet betray no qualms when debtors stand to lose twenty-seven pennies.

As a matter of fact the relative value of the precious metals depends not on relative output but on relative scarcity, and scarcity in each case is a compound function, depending on demand as well as on supply. It is not due to any variation in the output of the two metals that their relative value has changed, but to legislative enactments which have immensely enhanced the demand for gold, and thereby rendered it, relatively to silver, much more scarce.

To show how unsubstantial are all the allegations about dishonest money and the inequity of remonetising silver, we have only to observe what occurred in 1890. In that year we had an object lesson, which brushes aside all these nebulous theories as to the present market price of silver being in any degree a test of its true value, when once more available, amongst the western nations, as legal tender for the payment of debts. In 1890 the Sherman Act was passed (X.) extending the purchases of silver by the States. For a while it was believed that the United States were likely to remonetise silver at a fixed ratio, without waiting for the concurrence of any European nations. And what was the result? Silver, which in 1888 and 1889 had averaged in value 3s. 6d. the ounce, rose steadily in value, and in August and September 1890 reached 4s. $6\frac{1}{2}$d. per ounce, or within $5\frac{1}{2}$d. of its old value of 5s. the ounce. There was no hocus-pocus about it, nothing but the ordinary higgling of the market. Yet the mere prospect of one single nation undertaking, without international agreement, to receive silver at a fixed ratio to gold, sent the price up to 4s. $6\frac{1}{2}$d. per ounce, which corresponds to a ratio of $17\frac{1}{4}$ to 1. Here was silver, by the operation of the market, actually back at $17\frac{1}{4}$ to 1. No person was injured by it. On the contrary a great impetus was given to trade. Nothing can prove more absolutely to any unprejudiced mind, that the present relative value of silver and gold is the result of legislative demonetising enactments. The relative market value of silver ($17\frac{1}{4}$) in 1890 was the result of no adventitious circumstances, it was simply the result of the unfettered action of the bullion merchants, who, knowing accurately the comparative relative demand for the two metals, and judging that the remonetisation of silver was at hand, put it up to near about what they deemed to be its true value, if used as legal tender concurrently with gold. Had the United States carried out their intentions and gone through with the matter, the question might have been solved, but no one can blame them for hesitating single-handed to attempt to remonetise silver at a fixed ratio.

Take again the closing of the Indian mints in 1893, and what better object lesson could bimetallists desire? Within a fortnight of the closing of those mints silver fell 25 per cent in value. Could anything prove more conclusively the direct effect of legislative enactments upon the value of the monetary metals, and how wide of the mark is the monometallic view that legislative enactments are futile, as they can exercise no such control?

What would be the value of gold if the western nations were to cease coining it and were to adopt silver only as their standard of currency? The fall of gold would be relatively far greater than the fall in silver has been. It is the cessation of the demand for silver as a standard of currency which has lowered its market value, and with the resumption of that demand its value would return. That this is no mere surmise is proved beyond cavil by its rise in 1890, and in the opposite way the truth of this principle is further demonstrated, by the fall which took place in silver when the Indian mints were shut in 1893. Can any one suppose that if the English legislature were to decree that tall silk hats should no longer be worn, the stock of hats unsold would retain their present value? They would not preserve any value as articles of intrinsic artistic beauty. There can be no doubt they would cease to carry any value whatever. On the other hand, if such an enactment were repealed they would recover their value. It is precisely the same with silver. Once recreate the demand for it and the value, as in 1890, returns.

This brings us to the other assertion made by Lord Farrer, that English bimetallists desire to re-establish the ratio of 16 to 1. English bimetallists desire that a ratio should be re-established, but they have no desire that such a ratio should be other than the fair relative value of the two metals, and they deem it most unwise to dogmatise as to that value. All knowledge is by comparison, and all we can do is to seek guidance from the actual experience of the past. If the mints are opened again to the free coinage of silver, the value of silver will be very different from what it is while the mints are shut. The rise in the price of silver in 1890, when it was anticipated that the American mint would reopen, proves beyond a doubt that we have only to open the mints to silver, and that immediately silver will again be worth in gold more nearly its old than its present value.

Given a fixed ratio, there is no difficulty in maintaining it, as we have seen, although the output of the two metals may vary from time to time very considerably from it. If the Latin Union had not abandoned the fixed ratio, there would not have been any difficulty in continuing the old ratio of $15\frac{1}{2}$ to 1. This is evident from the fact that for half a century, from 1781 to 1830, the output of silver was not $15\frac{1}{2}$ to 1 of gold but 46 to 1. Yet during the whole of that time there was not the smallest difficulty in maintaining the ratio of $15\frac{1}{2}$ to 1.

Now, however, that the ratio for the time is gone, in re-establishing a ratio we should look at all the conditions of production of the two metals, and fix it as nearly as we can estimate the output, judging by the past and probable future production. It is just as in the case of a landlord letting a house to a tenant. As long as the tenant cares to remain, neither may wish to readjust the rent. But if the tenant quits and it comes to a new letting altogether, the landlord will look into all the circumstances, and may take the opportunity of a break in the tenancy, to revise the rent. English bimetallists have accordingly a very definite plan, and as to the ratio, they leave that to be settled at an International conference. For English bimetallists consider that, as the purpose of a fixed ratio is to provide the world with an International currency, such a currency should be settled by International agreement. What the exact ratio may be, must be left to such a conference to determine.

Before going into any conference for the purpose of settling a ratio between the two metals, a general understanding would have to be come to diplomatically beforehand, and in settling the limits within which the delegates to such a conference might act, many considerations would have to be taken into account.

It would be impossible to come to an understanding, unless it was made part of such an agreement, that the Indian mints should be reopened to the free coinage of silver. It would be of the first importance in settling a ratio to fix upon one which would not unduly disturb trade and existing contracts. True, statesmanship would strike a ratio which, taking into account all the existing equities, would not cause too violent a change in prices, but on the contrary lead to a gradual rise extending over several years. If it be said that many would suffer from a small gradual rise in prices, we have only to observe the period which opened in 1850 and closed in 1873, and ask where in those years were large classes injured. On the contrary, the writings of Mr. Newmarch and others (see XVI.) show that in this period of gradually rising prices, general prosperity prevailed.

Now and again suggestions are put forward that a fixed ratio should be re-established, and the ratio fixed somewhere about the present gold value of silver, say at 28 to 1 or 30 to 1. Those who make these suggestions are led to put them forward from having realised the serious evils resulting from the loss of a common currency. But they have not grasped the change that must ensue to the gold value of silver immediately Western Europe and the United States are prepared to reinstate it as legal tender. It is quite possible that a ratio of 20 to 1 might fairly meet all the equities, which, while providing the whole world once more with a common currency, would bring about gradually a general improvement in prices.

XIX

"BIMETALLISM WOULD DOUBLE PRICES"

MANY who "oppose bimetallism" give, as a reason for doing so, that "the re-establishment of a fixed ratio would double prices." Bimetallism would undoubtedly raise prices, but the rise would be gradual, and the *doubling* of prices would only be possible in the case of products from silver countries, which come to us almost wholly in an unmanufactured state. But let a monometallist state his own case—the case of a poor widow: "Suppose a widow with £300 per annum derived from £10,000 railway debenture stock upon which she now lives. Suppose the quantity of money doubled by bimetallism ; then commodities would be doubled in price, which to the widow would mean that £300 under bimetallism would only go as far as £150 per annum at present." But how is the rise in the value of silver (as compared with gold) to *double* the value of commodities other than those produced in silver countries? The great bulk of all goods imported into England from silver countries is imported in the raw state. It is raw products from silver countries, therefore, that will mainly be affected by a change in the value of silver.

The remonetisation of silver will gradually raise the price of wool, say, from 11d. to 1s. 6d. per pound. The "poor widow" wears a woollen dress. It contains, say, 3 lbs. of wool, and the cost of it when delivered by the dressmaker is now £2 : 2s. Suppose that with bimetallism the cost of the raw wool rises 7d. per pound, and the poor widow has to pay £2 : 3 : 9 for her dress, instead of £2 : 2s. Being a poor widow, with small means, she buys, say, a tin kettle which costs her now 2s. 6d. The tin in the kettle costs now, say, 3d., with tin at £60 per ton, and when, under bimetallism, tin rises, the kettle will cost 2s. 9d.

In the illustration given, our monometallist has to "suppose the quantity of money doubled by bimetallism." This could not possibly be; for it is impossible to obtain sufficient silver to double the quantity of money. When bimetallism is re-established it will only mean that whereas two persons out of every three now living in the world have silver for their currency, the third person also will use silver as currency in conjunction with gold. This will lessen the demand for gold and increase the demand for silver, and tend to draw the two metals nearer to their former relative values.

In the illustration given above of the evils of bimetallism, it is assumed that the "poor widow" had her capital invested in English railway debenture stocks. But suppose that some of her money was invested in consols. In this case she would have lost one month's interest annually after 1888 (receiving $2\frac{3}{4}$ per cent instead of 3 per cent), and in 1903 she will have another month's interest taken off. Suppose again that some of her money was invested in South American or Canadian railways, or in some foreign bonds, it is hardly possible but that she would have suffered some diminution of income, arising from the financial disturbances of recent years, due to the break in value between silver and gold. How far the remonetisation of silver will affect prices, must depend upon the ratio fixed between gold and silver. If the old ratio were to be re-established, prices would no doubt rise, and probably rise rapidly, but they would not rise to what they were prior to 1873, as improvements in manufacture will in the case of many commodities, account for a portion of the fall in values. If, however, a different ratio were adopted, say 20 to 1, it is probable that prices would only very gradually rise, and that the rise would be spread over a considerable period of time. But the first thing is to agree upon the principle of fixing a ratio. The point at which the exact ratio should be fixed, is then a question to be decided, after a full consideration of all the equities.

For the purpose of estimating the accuracy of the statement that "bimetallism would double prices," let us suppose that silver is remonetised at 16 to 1, as proposed by the Democratic party in the United States. It would be quite a mistake to think, that if this were done, "prices would be doubled." There are at least three things, which have to be taken into account, all of which would operate to prevent such a result.

In the first place, the ratio of 16 to 1 could not double the supply of the precious metals. It would only double the value of silver now current at market rates. Secondly, the increased demand for silver would lessen the demand for gold, and reduce its purchasing power. Thirdly, it has always been observed, when the metallic currency becomes more abundant, the rise of prices which follows, gives such a stimulus to trade, that a large portion of the increased supply of currency is absorbed in new undertakings before it can operate to enhance prices. The idea, moreover, that any change of prices, consequent upon an increased supply of metallic currency, would be instantaneous, is mistaken. Such a rise would in any event be a matter of time, though of course such a rise would be more rapid, if the ratio were fixed at 16 to 1, than it would be if fixed at 20 to 1.

XX

"BIMETALLISM IS PROTECTION"

THE notion that "bimetallism is a form of protection" is another fallacy arrived at by the process of false reasoning, and its genesis seems to be as follows. "Free trade lowered the price of corn. Bimetallism would probably raise the price of corn. Therefore bimetallism is protection." This is one of the fallacies of confusion, and assumes that because the result is the same, the nature of the causes are identical. We might as well reason that because men die both of heat and of cold, therefore heat and cold are one and the same, whereas they are the exact opposite of each other. If we are to apply the terms "protection" and "free trade" to bimetallism at all, it would be more correct to speak of the closing of the mints to silver as "protection" to gold.

Agriculture prospered well enough from 1850 to 1873. With open ports and open mints the English farmer did not complain. No protest was made then that the joint standard maintained by the Latin Union spelled protection for farmers. Let us see why it is that, as regards wheat and certain other products, the price has declined as the purchasing power of gold and silver has altered. A protective duty upon wheat, puts an extra charge upon all wheat coming into the United Kingdom from foreign ports, and to the amount of the duty charged, enables home producers to raise their price for their own products. The abolition of the protective duties on corn accordingly lowered the price of wheat, for the home producers could not get a higher price for their corn than imported corn fetched. But how has the change in the value of silver, as compared with gold, operated to affect the price of corn? It is true that the price of wheat has further fallen as a consequence, but not by taking away a tax which operated to enable home producers to demand a privileged price; it has acted in an entirely different way, namely as a bounty on exports from wheat-growing silver countries. And "bounties on export" and "protective duties" are alike indefensible.

According to the extraordinary reasoning of the monometallists, the abolition of a "bounty on exports," is precisely the same thing as imposing a "protective duty," and equally reprehensible. Bimetallism, however, instead of being a form of "protection," carries out on the contrary the principles of free trade. It would, on the one hand, open the mints freely to silver, and no longer adventitiously enhance the value of gold; on the other hand it would abolish what are now in effect "bounties on export" upon produce from countries where labour is paid in silver and not in gold.

That it should be brought as a complaint against bimetallism that it fosters protection is very strange, considering that the loss of the fixed ratio has done more than anything else to check the growth of the free-trade principle, and to promote the revival of protectionist ideas. It seems hardly too much to say, that unless bimetallism is restored, the free-trade battle will have to be fought all over again.

With open ports and—by the favour of the Latin Union—open mints, the prosperity of this country was unbounded, and so long as these conditions were maintained, no one in England thought of challenging the policy of free trade, while in foreign countries the principles of free trade were making steady headway. But since the loss of the joint standard a tide of protectionist reaction has set in. Germany, who pronounced for free trade before England did, has piled duty upon duty. The United States has a tariff bristling with protectionist restrictions. France, who was heading for free trade, is now again resolutely protectionist. In England, under the term "fair trade," protectionism has also made its presence felt.

The effect of the loss upon free trade of the old ratio of value between silver and gold, must be obvious if we reflect upon it. Free trade has to do with the unfettered interchange of commodities between nation and nation, and the settlement of commercial debts contracted between merchants dwelling within different geographical and political limits. So long as the exchange value of silver and gold remained constant, as was the case before 1873, the two metals in effect acted as one measure of value through all lands. The rise or fall of values of any specific commodities accordingly could only come about, on the one hand, from the increase or lessening of the cost of producing or of placing them on the market, or, on the other hand, from the contraction or expansion of the joint currency of both metals. In the latter alternative, commodities in all countries with a metallic currency would be equally affected.

But since the change in the relative value of gold and silver, resulting from the loss of the joint standard, a new element of variation in values has come in, bearing no relation to any decrease or increase of the cost of producing or of bringing commodities to market, and not affecting commodities equally in all countries, as was the case under the joint standard, when there was a decrease or increase of the metallic currency. The new element of variation arises from the change in relative value of silver and gold. And, as a result, foreign gold-using countries have in recent years become more and more protectionist, in order to protect their labourers from being driven from their employments by the cheap labour of the silver-using countries.

As we have said above, if the change in the relative value of silver and gold were of a *bona fide* character, and really due to a great and permanent change in the relative output of the two metals, it would justify the alteration of the old ratio, but would not justify the abandonment of the joint standard. But all the evidence is the other way, and goes to prove that the present market value of silver, as compared with gold, is entirely artificial, and the result of legislative causes. The restoration of the joint standard would give renewed vigour to free-trade principles.

The effect of "paper money" prices upon wheat is specially referred to later, Sec. XXXI.

XXI

"A FIXED RATIO CANNOT BE MAINTAINED"

WE have already explained the fallacy contained in the statement that a fixed ratio of value between gold and silver cannot be maintained (IV.), and how past experience has disproved this assumption. As this fallacy constantly recurs, we return to its further consideration, in order the more thoroughly to dispel this illusion.

The two tables which we give here, prepared for the Brussels Monetary Conference held in 1892, by Sir Guilford Molesworth, are full of instruction and afford ocular demonstration of the fallacy contained in the "cant" utterance about the impossibility of maintaining a ratio.

These tables have to be studied together, as the one illustrates the other. Looked at together, it is seen from the first table, how absolutely steady to the ratio, the market value of silver to gold remained, during the first seventy-three years of this century, and how after that time the relative market value of the two metals changed as much as to $34\frac{1}{2}$ to 1. It is seen by the second table that in the first two quarters of the century the output of gold, taken in value at

1 of gold to 15½ of silver, did not nearly come up to the output of silver; while in the third quarter of the century (1850-1875), the excess of gold over silver was even greater than the excess of silver over gold in the first two quarters of the century taken together. Yet in spite of these immense variations

TABLE XI

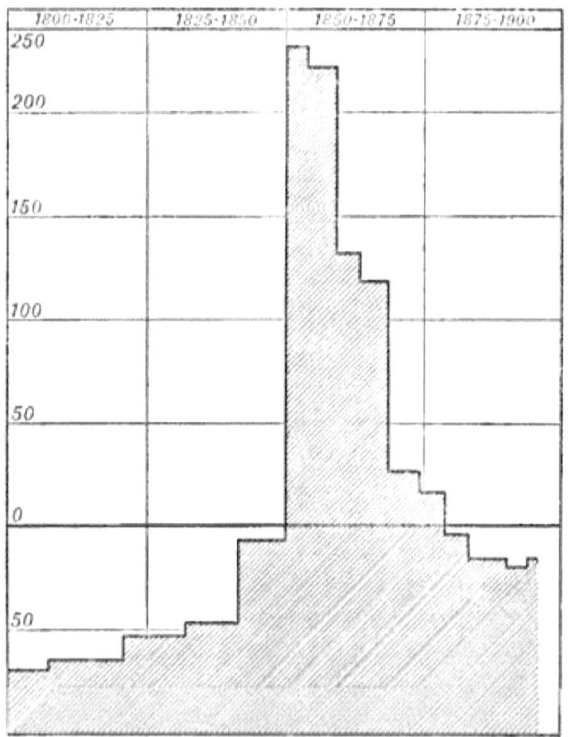

in the output, the ratio was steadily maintained, as Table XII shows.

When we pass the year 1873, after the fixed ratio is abandoned, although there is only a very moderate increase in the output of silver, an increase which will not bear any comparison with the excess of silver over gold in the first two quarters of the century, or with the excess of gold over silver in the third quarter; nevertheless, the relative value of the two metals is no longer maintained, and flies apart, quite out of all relation to the slightly in-

TABLE XII

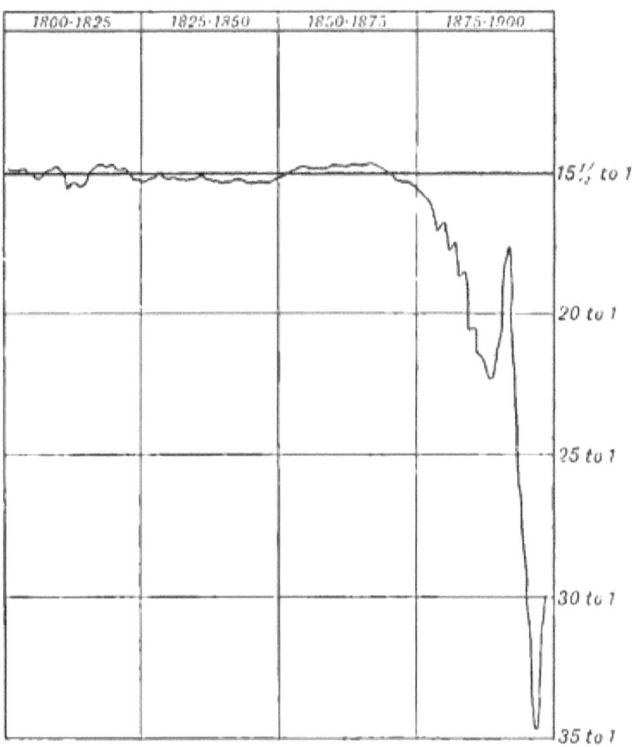

creased output of the white metal. To any impartial mind it must be obvious from the tables, (1) that a fixed ratio can be maintained, and (2) that the change in the relative value of the two metals in recent years is not due to variation in the supply. The only logical conclusion is, that the abandonment of the fixed ratio, and this alone, has caused the immense disturbance in the relative value of the two metals. The trifling fluctuations in the value of silver observable in the first three quarters of the century (Table XII.), are due to cost of transit either to or from foreign countries, as from time to time there was a deficiency in the supply or a superabundance in the London market. The extremes of variation under the fixed ratio were 4s. 10½d. per ounce and 5s. 2¾d. per ounce, showing a variation not at any time exceeding 3½ per cent on either side of the central point of the fixed ratio.

The way in which the fixed ratio was maintained in the years 1849 to 1856, is proof positive, that a fixed ratio can be maintained, and maintained during large fluctuations in the relative output of the two metals. In the year 1849 the annual output of gold was about £5,400,000, while the output of silver, valued at 15½ to 1, was £7,800,000. In 1856 the output of gold—owing to California and Australia—had risen to about £29,500,000, while the output of silver had remained almost stationary. Yet this sudden and considerable increase did not disturb the ratio, and the relative value of the metals remained the same in 1856 as in 1849. After such an experience it is clear that the fixed ratio can stand, without alteration, great changes in the relative annual output of the two metals. The real reason of this is no doubt due to the fact that, be the annual output what it may of either metal, the output bears so small a proportion to the existing stocks of the two metals in the world, that it cannot appreciably affect them, except over a very long period of time. Men who are supposed to be authorities upon these subjects, speak as if gold and silver were like apples and potatoes, and that for the supply of the precious metals, we depended upon the annual crop. This is one of the fallacies that we must put away, if we intend to understand this question.

Here we may with advantage look at the following table, and note the growth of our foreign trade.

TABLE XIII

EXPORTS AND IMPORTS OF UNITED KINGDOM

	Exports of British Produce.	Total Imports and Exports.	Silver per ounce.
1832	£38,250,000		5s.
1850	71,360,000		5s.
1860	135,890,000	£375,000,000	5s.
1870	199,580,000	547,380,000	5s.
1873	255,160,000	632,290,000	5s.
1880	223,060,000	697,640,000	4s. 4d.
1889	248,930,000	743,230,000	3s. 7d.
1890	263,530,000	748,940,000	4s.
1891	247,230,000	744,550,000	3s. 9d.
1895	226,169,000	702,820,000	2s. 6d.

In the above table the three years 1889, 1890, 1891, have been given as they show how in 1890, trade responded to the rise in the gold value of silver. In August and September of that year silver, as we have seen (XVIII.), rose to 4s. 6½d. per ounce, and this gave silver countries a greater power of purchasing in the English markets.

XXII

"ANY ALTERATION OF AN EXISTING STANDARD OF VALUE IS AN EVIL"

THOSE who "oppose bimetallism" give as one of their reasons, that "any alteration of an existing standard of value is an evil" (Lord Farrer). This is one more of the many fallacies with which the monometallists overlay the restitution of the world's common currency. *It is seen* that any change in an existing standard of currency is an evil. This every one is ready to assent to. *It is not seen* that the existing standard of value is subject to continual alterations, and that, judged by the fourth canon of a sound currency—stability of value (XII.)—the existing standard is lacking in an essential element of a good standard of currency.

It is not seen that the purpose bimetallists have in view, is the putting an end to the continual alterations in the existing standard of value. So far from desiring to *alter* the existing standard of value, they want to put a stop to its continually changing value, and a main purpose of bimetallism is to return to a stable standard of currency which will not be continually altering in purchasing power. "If I want to do business with a country with a different language," writes Dr. Süess, "I can learn the language, and the difficulty has disappeared. If I have to use another monetary unit I can calculate its equivalent, and the intercourse is free and fair; but if abroad all labour and all goods are calculated, not in a different monetary unit, but in a different metal, and if the value of these two metals varies, a fair intercourse is scarcely possible."

"The question stands: Whether it is better to tolerate all the injustice, all the chances, all the poisoning uncertainties of the present situation for another period, and then face a great crisis, or settle the question quickly, and give to the commerce of the world peace and a common equal basis of the exchange of all commodities."

In short, the "alteration in the existing standard of value" which monometallists deprecate, accurately describes the condition of things which *now* exists, and this bimetallists seek to put an end to. The monometallists who would let things drift on as they are doing at present, are those who would perpetuate "alterations in the existing standard of value."

XXIII

"BIMETALLISM WOULD BE A DANGEROUS EXPERIMENT, AND A THROWING OF THE SOVEREIGN INTO THE CRUCIBLE"

THE assertion that "bimetallism would be a dangerous experiment and a throwing of the sovereign into the crucible," is a fallacy which can mislead us, only so long as we turn a deaf ear to monetary history. An experiment is an act or operation undertaken to discover something unknown. The joint standard, or bimetallism, as we have seen, stood the test of centuries, under conditions as severe as any that are ever likely to arise. So a return to bimetallism could be no experiment at all. If the term experiment is to be used, it would be more applicable to the period of currency unrest which we have experienced since 1873. Since that year, for the first time in the world's history, we have entered upon a period with the link between gold and silver broken, and this experiment—this trial of the unknown—has discovered for us a new phase of continual alterations in the purchasing value of gold, in other words, of marked instability in the existing gold standard of currency.

The term "experiment," is of the nature of the "no innovation" or "hobgoblin" fallacies. But, as we have shown, it is singularly inappropriate in the present case, as bimetallists have no desire to venture upon any experiment, nor to throw the English sovereign into the crucible; their desire is rather to take the sovereign out of the crucible, and to return as speedily as may be to a safe and well-beaten track. Speaking of bimetallism as "an experiment," leads people to forget that bimetallism has a history, and a history that illustrates and justifies every argument used by bimetallists. But the history of bimetallism the monometallists fight shy of, and they seek to throw dust in the eyes of the unwary, by speaking of bimetallism as an "experiment" of which "no one can see the ultimate result."

Put about by the difficulty of meeting the historical argument in favour of a joint standard, monometallists have resorted to the device of trying to differentiate between a new and an old bimetallism, saying, that "the new (international) bimetallism is a wholly new thing both in theory and practice." The system of Locke and Newton was an open mint for both metals as legal tender, at a ratio fixed by law. That is the principle of modern bimetallists. Does it become a "wholly new thing both in theory and practice," by the mere fact of other nations agreeing to the principle in every detail? The only thing that is new about bimetallism is the name "bimetallism," but the term "monometallism" is equally new. The allegation that the new bimetallism is wholly new both in theory and practice, is satisfactory in so far that it shows to what desperate plights monometallists are driven in "opposing bimetallism."

Professor Foxwell has justly observed, that our present policy is not only an experiment, but one calculated to lead us directly to other monetary experiments of the most doubtful character. "The true remedy for the existing monetary anarchy is remonetisation of silver. The alternative policy is one of drift. It is more than probable that the evils of monometallism will force on a series of doubtful experiments, from which England will be the first to suffer, to one of which she has already been driven in closing the Indian mints. We may refuse to have a share in determining the world's monetary policy, but we cannot avoid being vitally affected by it. England is of all nations the most interested in securing a sound settlement of the international monetary question, and there is no country upon which rests the duty so clearly, of taking a leading part in bringing about this great international monetary reform."

The closing of the Indian mints has not been the only experiment, for the United States passed the Sherman and Bland Acts (X.), with the object of mitigating the rigour of the contracted gold currency, both of them monetary experiments ill calculated to lead to the only true solution, namely, the remonetisation of silver.

It is generally allowed when results are accurately predicted, that the reasons given for such results thereby receive confirmation. In the present controversy it should be remembered, that when the anti-silver crusade took place in 1873, the consequences of the demonetisation of silver were accurately predicted, before the results of that policy were felt in practice. "Nothing is more instructive," writes Professor Foxwell, " than to compare the forecasts of such men as Ernest Seyd, Wolowski, Baron Alphonse de Rothschild, and the bimetallists generally, with the ablest of the monometallist writers."

XXIV

"BIMETALLISM WOULD MAKE ENGLAND A DUMPING-GROUND FOR SILVER, AND DRIVE GOLD OUT OF CIRCULATION"

MONOMETALLISTS say, "if England coined both silver and gold, our country would become a dumping-ground for depreciated silver, and our gold would go elsewhere—the cheap metal would drive out the dear." Those who make use of this fallacy, have not troubled themselves to understand the position of English bimetallists. Such an assertion as to the effect of bimetallism has no meaning, unless we are to assume that England alone is to revert to the joint standard. But this is exactly what English bimetallists do not propose to do. The whole of their arguments have been directed to a "joint agreement with other countries," on a basis sufficiently wide, to ensure the stability of value which was enjoyed during the existence of the Latin Union. Those who imagine that one metal would drive out the other under a system of bimetallism, are thinking of National, not of International bimetallism. They are tilting against the national bimetallism approved by the Democratic party in the United States, not against the international bimetallism approved by the Republican party and European bimetallists.

With a "joint agreement" there is no possibility of one country becoming the dumping-ground of either metal, for the simple reason that no one would have anything to gain by it. Every one at his own mint—in Germany, France, or the United States—would be able to coin silver or gold, and meet all his obligations with either. Gold would have no greater paying power than silver, and if any were so foolish as to put themselves to the trouble of "dumping their silver" say, in England, it would be equally open to us, under the joint agreement, to do precisely the same thing in return. But the argument cannot be treated seriously, as it rests upon the entirely erroneous assumption, that we are desirous of opening our mint to the free coinage of silver, without first entering into an agreement with any other countries to do the same, and to adopt an identical ratio.

Some have objected to an "international" agreement on the ground that it would to some extent deprive us of our "independence." The same argument would hold good against the International Postal Convention, but no reasonable man would wish to abandon it, for a visionary "independence."

The underlying idea of the "dumping-ground" fallacy is, that one metal would be cheaper than the other. "If," says a leading monometallist, "the payer has an option to pay in one of two metals, and the option has any meaning, he will of course pay in the metal which is cheapest." The notion of one metal being cheaper than another, is not applicable to a bimetallic standard, for in a bimetallic country gold money cannot be dearer than silver money, nor silver money dearer than gold money, since gold coins and silver coins can only possess by law but one value, and no question of dearness or cheapness can arise between them. Like a compensated pendulum, the joint-standard would swing under all conditions with an even beat.

The value of both metals as currency depends upon their monetary use (IV.). To say, as the Gold Defence Association states, that the value of one commodity in terms of another cannot be fixed by law, is economically true, but to apply this principle as they would do, to the joint monetary standard, is to forget or to fail to understand, that gold and silver used as money are not commodities, and that when used as a joint measure of value their values are not independent of each other.

XXV

"IF THE MINTS WERE REOPENED TO SILVER, EVERY ONE WITH A BALANCE WOULD DRAW IT OUT IN GOLD AND EVERY BANK WOULD BREAK"

It is somewhat hard to understand how any persons can bring themselves to believe, that if the mints were reopened to silver, "every man and woman who had a balance in the bank would straightway go and demand it then and there in gold, and at the end of a week not a bank in the United Kingdom would be left standing." Still this has been put forward in apparent good faith.

The answer to this statement is, that it is not only untrue, but diametrically opposed to the experience of everyday life. For it is a well-established fact, that the public become buyers in rising and not in falling markets. The statement we have quoted, is made upon the assumption, that the ratio to be fixed between the silver and gold would not be 30 to 1, but some other ratio nearer to the old ratio of $15\frac{1}{2}$ to 1. Assuming for this purpose, that the ratio were to be fixed at 20 to 1, then we are asked to assent to the proposition, that all our previous experience of the ways of the public would in this case prove no guide, and that the public, instead of seeking to secure the metal rising in value, would exert themselves to obtain the metal which, relatively to the other metal, would be falling in price.

As a matter of common experience, very few of the ordinary public trouble themselves about the rise and fall of values sufficiently, to sell or purchase with a view to obtain a profit. And if they did, the movement in the relative value of the two metals would be so stealthy, that they would have little opportunity of profiting by it, even if they made the attempt. But instead of being anxious to obtain gold, as the monometallists say they would be, we should find them on the contrary anxious to obtain silver as likely to yield them a profit.

When silver rose in 1890 to 4s. $6\frac{1}{2}$d. an ounce —within $5\frac{1}{2}$d. of its old value of 5s.—it crept gradually up from week to week, as the chances of the reopening of the American mints seemed to improve. When the western nations of Europe agree again to remint silver upon a fixed ratio, a similar gradual enhancement of the price of silver will be observed, and before the public think of making any move either to lock away gold—as the monometallists would have us believe—or to possess themselves of silver—which is far more likely of the two—silver will be within 6d. of its old value, as in 1890, and it will be worth the while of no one, either to sell gold or buy silver in any quantity.

We have had a precisely parallel experience in recent years in the Argentine Republic. When, after the Baring troubles, gold went to a high premium in that country, and paper money depreciated, the Argentine speculative public, true to all experience, everywhere purchased the rising metal. The more gold rose, the more they were inclined to buy it and lock it away. The Government used every endeavour to keep down the premium on paper money, but without avail. In those years, when the premium on paper was very high, more and more gold was absorbed by the public purchasing the rising metal. The following statistics of the imports and exports of gold into the Argentine Republic in recent years confirms this.

TABLE XIV

GOLD IMPORTS AND EXPORTS

ARGENTINE REPUBLIC

	Imports.	Exports.
1890	£1,389,000	£1,000,000
1891	1,777,000	236,000
1892	1,244,000	295,000
1893	877,000	103,000

I have no later figures, but these are sufficiently striking. In 1890 the gold premium—after hovering between 40 and 80—took its first leap to 160, and in the next year (1891) another leap to 280. As the premium rose, the balance of imports over exports of gold rises in a marked degree. Instead of the country exporting gold to meet its foreign obligations, more gold on balance of imports and exports went into the country, and the public absorbed it. The reason was the one we have given, that the public always buy on a rising market, and gold was the only commodity which was then rising in the Argentine Republic, so it was bought. For the very same reason, when the premium on gold began to decline, it fell more rapidly than many anticipated, the reason being again the same, that the public came forward to purchase with gold the paper money, which in its turn was rising.

The same thing has been experienced in England and in other countries. After the great war with Napoleon, when gold was at a premium, the Government made arrangements for gradually taking up the paper currency at par. But long before the date they had fixed for the resumption of specie payments, paper was practically back at par, the reason being, that as soon as the matter was definitely settled, those who wanted to buy for a rise naturally bought paper and thereby brought it back to par by anticipation. The same thing occurred in the United States when specie payments were resumed after the Civil War. As bimetallism would enhance the price of silver, it would be silver that the public would rush to buy and not gold.

The Honble. R. B. Mahany has elaborated this fallacy in the *North American Review* (1896). If the mints are again opened to silver at a fixed ratio, he prophesies, that "depositors in banks would demand payment in gold. Runs on these institutions would cause half of them to close. Employers doing business on credit would fail. Working men would be thrown out of employment. Crash and panic would be the continuous order of the hour."

When those who would instruct the world, flounder with the utmost assurance into such absurdities, it is evident that the spirit is still abroad which condemned Kepler as a dreamer, and prevented Copernicus for thirty years from publishing his treatise on the Heavenly Bodies. "We Bees do not provide honey for ourselves!" exclaimed Van Helmont, the father of modern chemistry, worn out by the laugh of incredulity. But fallacies wither away and dry up when exposed to the steady light of human knowledge.

XXVI

"THE SUPPLY OF GOLD COMING FORWARD FROM THE TRANSVAAL AND ELSEWHERE WILL SOLVE THE QUESTION"

A PROMINENT official of the Gold Standard Defence Association, writing to the Bristol Chamber of Commerce (29th Oct. 1895) said, that the fortunes of Bimetallism were sealed, because "the large output of gold in South Africa has practically settled the question." This statement is a curious and useful admission of the soundness of the Bimetallic contention, that the scarcity of gold is responsible for the decline in prices and the consequent depression of trade. The admission that an increased output of gold would be a cure for present troubles, concedes the truth of all that Bimetallists have said as to low prices being the result of a constricted currency.

But is there any reason to suppose, that gold in sufficient quantity, to redress the balance of our loss from the demonetisation of silver in the western world, is likely to be forthcoming. "The older countries are exhausted, and to-day we are raking together with the most perfect machinery what remains in distant parts of the world. The result is a considerable temporary increase in the yearly production, but means a quicker exhaustion of the supplies."

For a quarter of a century (1851-1873) the world's monetary metals were jointly increasing, to an average amount of £36,000,000 annually. When in 1873 the western nations discarded silver, they deprived themselves forthwith of half the annual output of the world's metallic currency.

Since 1873 the production of gold has amounted to about £500,000,000. Of this amount fully two-thirds is required for the arts, leaving about £170,000,000 to be added to the world's stock of gold money. An estimate, derived from published sources of information, computes the additional demand for gold by the world's mints, owing to the demonetisation of silver in 1873, at £400,000,000. So that after all these years £230,000,000 is still needed. (See Lord Aldenham's *Colloquy on Currency*, p. 115.)

This estimate of the amount of gold available for the purposes of currency is a very liberal one, as it leaves out of account the demand of India for gold for non-monetary purposes, which is very considerable. The better opinion seems to be that only one-fourth of the annual product of gold is available for currency purposes, owing to the great wear and tear of gold and its vast consumption for artistic purposes.

As an illustration of the extreme sensitiveness of the money market, resulting from carrying on the trade of the Western world upon the narrow basis of a gold standard only, we may refer to the effect upon the market values of securities, of the withdrawal of £12,000,000 from the Bank of England, between the first week of July and the third week of October 1896. In the middle of August, the aggregate value of 325 representative securities dealt in on the London Stock Exchange, was £3,271,000,000. On the 20th October, the same securities were valued at £3,156,000,000, showing a loss in eight weeks of £115,000,000 (see *Bankers' Magazine*). It is impossible to suppose that, under a bimetallic system, the withdrawal of twelve millions sterling would have anything approaching the same effect upon the general market value of securities. As long as all is dependent upon the one metal, the world's chief market must, under the most favourable circumstances, be in a condition of unstable equilibrium.

XXVII

"A SINGLE STANDARD OF CURRENCY IS BETTER THAN A DOUBLE STANDARD"

WE commenced this little treatise by showing, that those who call themselves monometallists, advocate principles which absolutely debar the world from enjoying the endless benefits of a common currency. Monometallism means two separate standards of currency, ever varying the one from the other. In defence of this barbarous condition of things, which throws gold and silver countries, in their international dealings, back upon the rudimentary system of barter, monometallists plead that "a single standard of value is better than a double standard of value." This axiom, borrowed from economic writers, is one which no bimetallist would wish to dispute, but, used as the monometallists use it, it misleads the unthinking, who do not perceive the fallacy contained in the application of this axiom by the monometallists.

"That a single standard of value is better than a double standard of value," is one of those propositions which commends itself to the mind at once. The advantage of a single metal as a measure of value of all commodities needs no justification. *It is that which is seen.*

It is not seen that the world has no "single standard of value," and that there is no likelihood of its ever possessing one, unless the whole world adopted silver as its standard of value, which is very improbable. There is not enough gold as it is, to go round for a third of the population of the world, so that the idea that it is within the bounds of possibility that gold can become the "single standard of value" may be dismissed as idle.

It is not seen that those who approve of bi-metallism do not advocate a double standard, but a joint standard. *It is not seen* that the economists who have said that "a single standard of currency is better than a double standard," have never laid it down, that "no common standard of currency is better than a joint standard." It is no use for monometallists to imagine that the axiom "that one standard of currency is better than two standards of currency," settles the question in their favour. Bi-metallists too, give ready assent to the same axiom, but they reasonably ask where is this "single standard." To this there is no response, for no such standard exists or can exist, unless we adopt the bimetallic principle of a joint standard.

XXVIII

"IF BIMETALLISM RAISED THE PRICE OF SILVER, THE GAIN OF THE SILVER USERS WOULD BE THE LOSS OF THE GOLD USERS"

THERE is, as we have seen, every reason to think that the reopening of the mints to silver would raise its price measured in gold. For the joint use once more of silver along with gold, as legal tender, would create such a demand for silver, as would raise its gold value approximately to the ratio value so long maintained prior to 1873. "Then," say the gold monometallists, "assuming that this change in the gold value of silver would ensue from the reopening of the mints, one man's gain would be another man's loss, and silver-using countries would gain at the expense of England, which is a gold-using country."

The fallacy of this argument lies in the assumption that one man's gain *must* be another man's loss. There are gains which are losses to no man, just as there are losses which are gains to no man. The common idea that one man's loss is another man's gain, comes from gaming and wagering, and in such cases what is one man's gain is strictly another man's loss. The measure of the loss of the one is the measure of the gain of the other. One loses and one gains. But there are other losses.

The clumsy son of the grocer breaks the shop window, and his father has to pay the glazier £5 to mend it. *It is seen* that the grocer loses £5 and the glazier receives £5. *It is not seen* that the gain of the glazier is the loss of some other tradesman. For the grocer having to spend £5 on mending the broken window has £5 less to spend on other things. In the case of the broken window the gain of the glazier is at the expense of other tradesmen, with whom the £5 expended upon new glass would otherwise have been spent, and in addition to their loss of custom to this amount, there is the loss of £5 to the grocer, who is no better off than he was before. In this case, therefore, two persons lose and one gains.

Next take the case of a farmer, whose rick of hay or of wheat is burnt to the ground; or that of the owner of a statue by Canova or of a picture by Reynolds, destroyed by accident or intention—these are losses which bring gain to no one.

On the other hand, there are innumerable gains which bring loss to no man. If we take a large catch of fish, or grow a heavy crop, discover a new mine, or invent a new mechanical contrivance, no single person need be one penny the worse. If we invest our money in a railway or some other company, which by good management is able to return to the shareholders double the interest it formerly earned, and the capital value of our stock is thereby doubled, no one is one penny the worse. A shareholder a few years ago might have purchased stock in the London and South-Western

Railway for £100, which to-day he can sell for £200. No one is a loser by this transaction; it is a gain at the expense of no one. And so if silver rises in value, it is a gain at the expense of no one. If we have a silver teapot and other articles of silver in the house, which collectively weigh 100 ozs., and to-day are worth £25, no one will be a loser if they rise in value and should become worth £37 : 10s. The owner gains, but no one loses.

It must be remembered, moreover, that of the silver passing current in the world as money, about one-third of it is in the pockets of subjects of the British Empire.

The phrase "one man's gain is another man's loss" is an equivocal expression which is only true under certain circumstances. Like a similar phrase, "one man's food is another man's poison," it is not to be taken as true under all circumstances, but as true only under particular circumstances.

Some monometallists have objected to the remonetisation of silver on the ground that the silver miners would profit by the change. "But no one," observed M. Allard, "has ever suggested that we should shiver in winter, lest coal miners should grow rich."

XXIX

"BIMETALLISM MEANS INFLATION OF THE CURRENCY"

In considering the assertion that bimetallism means "inflation" of the currency, we have to ask ourselves what is an inflated currency. An inflated currency is a currency which is based upon a supply of specie (coined money), or bullion (ingots), insufficient to meet the paper issued against it. The Bank of England is allowed to issue a certain number of Bank of England notes against a debt due to it from the English Government, but every additional note that it issues can only be issued against gold (specie or bullion) lodged in the issue department. In some countries, such as the Argentine Republic, there is no sufficient supply of gold held against the notes issued, to enable them to be paid in cash when presented. In such cases a new note will be issued in return for an old one, but specie or bullion will not be exchanged for it. When this is the case the note issue goes to a discount, and payment in cash can only be obtained by paying a premium. The currency is then inflated.

It is impossible therefore to use the term inflation, in regard to the bimetallic proposal to make silver, jointly with gold, legal tender. For a currency is inflated when the notes, which for purposes of convenience pass current in lieu of specie, are so numerous, that they can no longer be exchanged at par for the specie which they are supposed to represent—in other words, when the notes representing the standard currency of a country are in excess of the currency they represent. It cannot be applied to an addition to a standard metallic currency.

Inflation of the currency is universally condemned by all economic writers; so much so that condemnation of an inflated currency has passed into one of the axioms of everyday life. Fully aware of this, and acting on the principle of "giving a dog a bad name," monometallists have asserted that bimetallism spells inflation, whereas as a fact, it is an absolute impossibility for it to do anything of the kind. Specie or bullion is the substance and note issues are the shadow. When the shadow is excessive a currency is inflated, but by no process of reasoning can a currency become inflated when the substance, or, in other words, the standard legal money (specie or bullion) is added to in quantity.

No doubt there are some in the United States who support bimetallism who are inflationists. "They are those who in 1868 and 1876 were against the resumption of specie payments, and were in favour of prolonging the period of inflation. Foiled in their attempt to secure an inflated currency with irredeemable greenbacks, they support the cause of silver as giving an enlarged currency." But there is all the difference between an enlarged currency of the precious metals, and an inflated paper currency which is merely the product of the printing press. It is the difference, as we have said, between substance and shadow, between a real and unreal currency. Bimetallists therefore cannot be inflationists even if they would, for no increase of the actual standard of currency can be inflation. But when we come to consider why, apart from the primary use of a bimetallic standard (that of affording the world once more a common and a stable currency), bimetallists should support an increased currency, the reason is not far to seek. Though not inflationists they are *against the contraction* of the currency. The Royal Commission on Depression of Trade (1885) reported that there was "a continuous fall of prices caused by the appreciation of the standard of value," and all history teaches us, that from the time of the discovery of the Potosi silver mines, which enlarged the currency, down to the demonetisation of silver, which contracted the currency, commercial prosperity has followed with unfailing regularity an increased output of the precious metals, while commercial depression has followed with equal regularity any contraction in the currency. When the Roman Empire broke up, the metallic money then in use amounted to some £350,000,000. With the fall of the Roman Empire the output of the precious metals practically ceased. By the end of the fifteenth century the metallic money in use had sunk to £40,000,000. In 1455 the price of wheat per bushel was twopence (Adam Smith). With the contraction of the currency population, freedom, commerce, arts, and wealth had disappeared. The discovery of the new world restored the volume of the precious metals, brought with it rising prices, and vitalised trade and commerce. The fruits of material prosperity have been gathered in the marvellous spread of modern civilisation.

XXX

"THE GOLD STANDARD IS THE CAUSE OF ENGLAND'S PROSPERITY"

IT is difficult to understand how the idea came to be held, that "the cause of England's greatness is her gold standard." Nevertheless the belief has reached high places, for a Chancellor of the Exchequer has stated, that England has no intention of departing from its existing system of currency, "*upon which its unequalled prosperity has been founded.*"

This is a clear and definite assertion, and, if true, we should reasonably expect to find—

1. That after England adopted the sole gold standard and discarded silver in 1816, her commercial prosperity and influence increased in a marked degree.

2. That when the Latin Union and the United States abandoned silver in 1873, and completed the break with silver begun by England in 1816, the prosperity and influence of England would have still further increased.

3. That foreign countries possessing the sole gold standard, would share in the prosperity which England is alleged to enjoy, as a consequence of the gold standard.

With regard to the first point, there can be no dispute about two things. Firstly, that we had a joint standard of currency in England down to 1816. So that if the single gold standard is the cause of England's prosperity, that prosperity dates from a period subsequent to 1816. Secondly, those who introduced the change to the single gold standard, never even suggested that the abandonment of silver was going to inaugurate an era of prosperity for this country. They, at any rate, have left no indication of holding any such opinion. A reference to Mr. Sauerbeck's tables, moreover (see Appendix) shows, that the abandonment of silver inaugurated a prolonged period of commercial stagnation and disaster, a period which only terminated with the expansion of the currency after the Australian and Californian gold discoveries.

We have only to turn to the pages of history to see for ourselves what was the condition of England during the last century, when the joint standard was her currency. "By the middle of the century our export trade had increased to double what it had been when the century opened. With wealth and prosperity prices steadily rose. Twenty-four years later, England ceased to be a mere European power. Mistress of North America, dominant in India, and claiming as her own the empire of the seas, she suddenly towered above the nations of the continent. At the end of the century the progress made was wonderful. The population had more than doubled, and the advance of wealth was considerably greater than that of the population. The loss of America had increased the trade with that country, and England was already far on the road to become the workshop of the world." At the close of the French war, when we abandoned silver, what is the account of England? "During the earlier years of the war the increase of wealth had been enormous. That war gave England the possession of the colonies of Spain, of Holland, and of France. English exports had nearly doubled since 1800. At home there was a vast increase of capital and a vast increase of population, while abroad England was sole mistress of the seas."—(Green's History *passim*.)

If we push our inquiries further back, we find that as long ago as 1660, when the English currency was silver, and silver only, General Monk described England as "the Metropolis of Commerce."

Such, then, was the condition of England before she adopted the sole gold standard, which we are told has been "the cause of England's prosperity."

After the French war, when we resumed specie payments, upon the sole gold basis, English trade and commerce entered upon a prolonged period of stagnation and depression, only shaken off after 1850, when the gold currency expanded owing to fresh discoveries and supplies of the precious metal. From that time England prospered greatly, until the end of 1873, when we entered upon a fresh period of prolonged stagnation, worse than that encountered earlier in the century; a period which is contemporaneous with the time "when the support of the French law being withdrawn, England was left in perfect and unalloyed enjoyment of the sole gold standard."—(Lord Aldenham, *Nat. Rev.* 1896.)

Yet this was the very moment when, if England's prosperity was really due to the gold standard, we ought to have entered upon a period when trade and commerce would have shown exceptional growth and vigour.

The true cause of England's prosperity is of course due, not to her currency, but to the energy and enterprise of her people, which has enabled them to make England the chief market of the world. The enormous commercial business carried on by English traders has been fostered by the happy geographical position of the country, which, lying outside the area of continental wars, has afforded us more than two centuries of uninterrupted internal peace. Added to this we have been favoured by an unlimited store of coal and iron, and furnished with harbours unequalled for maritime trade. The energy of the people coupled with the advantages flowing from peace, position, and natural resources, has made England the greatest mart of the modern world, and her vast trade transactions has made London, as the centre of England's trade, the pivot of the world's money market.

If the sole gold standard were "the cause of our prosperity," then not only would England be commercially great, but so also would be Chili, Sweden, Norway, Peru, and Portugal, for these countries also enjoy, in its integrity, the sole gold standard. But the mere mention of these countries, as occupying a dominant commercial position, is sufficient to disprove any such title. If we turn to France, Germany, Austria, or the United States, great as these countries are, they cannot be cited as possessing the sole gold standard, for although they have closed their mints to the free coinage of silver, they have yet vast quantities of silver in circulation, which is still legal tender. Their demand, still unsatisfied, for gold to replace their silver currencies, is comprised in the estimate given of gold requirements in Sec. XXVI. Until these requirements are met—and there appears to be no prospect of this coming about for many years yet to come—these countries will not be in the enjoyment of the sole gold standard, but must be content with their present position, which is generally known as that of the "limping standard."

It would be far nearer the truth to say that the sole gold standard is imperilling England's commercial pre-eminence, than it is to say that the gold standard "has caused England's prosperity." But we have only ourselves to blame in this matter. The evil example we set in 1816 has led other nations to complete the break in the chain of the world's currency, to our immense detriment as a commercial nation. The *virus*, with which Lord Liverpool inoculated the world in 1816, after lying dormant until 1873, then took effect, in a way only too plain to those who will not shut their eyes nor stop their ears.

XXXI

"SOFT" MONEY AND PAPER CURRENCIES: EFFECT ON WHEAT PRICES

THE expression "soft" money is used to describe paper money in contradistinction to metallic—gold or silver—money, often called "hard" money. When paper money in supported by a due amount of metallic money in reserve, it is to all intents as good as metallic money. When this is the case, as in England, no one thinks of calling it "soft" money. It is only when the metallic reserve is insufficient, or is thought to be likely to become insufficient, that the term "soft" money is applied to a paper currency.

It is desirable to say a few words about "soft" money, as in certain ways the "soft" money of the Argentine Republic in recent years has intensified the fall in the price of wheat and wool. These commodities, unlike most others, have been subject to a double fall in price, one fall in price arising from the altered value of silver as measured in gold, the other fall in value arising from the still greater depreciation, as measured in gold, of the Argentine paper dollar.

From 1885 to 1890, wheat sold for about 32s. a quarter and had reached the limit of its fall so far as silver could affect it. Then came the Baring collapse, which disorganised finances in the Argentine Republic and sent their paper dollar down to 25 cents (1s.) gold. If the Argentine paper dollar had only fallen to 50 cents gold, it would have been upon a par with the Indian rupee as measured in gold, and wheat would not have dropped to lower prices, but the fall to 25 cents gold, further depreciated the price of wheat, and it fell to 21s. the quarter. This enabled the Argentine Republic to cut into the Indian wheat trade at lower prices.

In 1892 India exported 56,000,000 bushels of wheat and the Argentine Republic exported 18,000,000 bushels. In 1894, owing to the depreciated Argentine paper dollar, the two countries had reversed their position, and while the Indian export had fallen to 13,000,000 bushels, the Argentine export had risen to 59,000,000 bushels.

The gradual fall in the gold premium at Buenos Ayres is now raising the value of wheat, and this rise will continue until the premium on the paper dollar falls to 100, when the rise will cease. After that point the value of wheat—allowing for passing fluctuations due to failure of crops or abundant harvests—can rise no further until there is a rise in the gold value of silver. Wheat and wool are the only two commodities of importance which have been affected by the second fall in values arising from the depreciation of the Argentine paper dollar. The Argentine Republic has accordingly undersold and displaced India for a time and further lowered the price of wheat.

TABLE XV

THE WORLD'S OUTPUT OF WHEAT

FROM THE CORN TRADE YEAR-BOOK

In 1892, 305,170,000 qr., and the average price 30s. 0d.
,, 1893, 314,054,000 ,,   ,,   ,,   24s. 6d.
,, 1894, 320,365,000 ,,   ,,   ,,   22s. 6d.
,, 1895, 311,490,000 ,,   ,,   ,,   23s. 0d.

The explanation given of the fall in the price of wheat since 1891, is fully confirmed by published tables, which show that the fall in the price of wheat between 1891 and 1894 was not due to the increase of the output.

The above table shows a fall of 7s. a quarter, equal to a fall of 23 per cent, while the output has only increased just over 2 per cent. The earlier fall in the price of wheat was due to the drop in the value of silver as compared with gold. The more recent fall in the price of wheat—the heavy drop since 1891 —is due to the depreciated paper in the Argentine Republic.

Having reached this point, we may here with advantage reflect upon the marked differences which exist, between the "soft" paper money of the Argentine Republic, and the "hard" silver money of India. On the one hand there is no limit to the depreciation of a paper currency, as its supply is merely dependent upon a liberal use of the printing press. On the other hand, and this is very important, a depreciated currency brings every country indulging in it, into discredit. There may be times, under stress of war, when it cannot be avoided. But no country is proud of it, and every country feels that it casts a stigma upon it; that it is only endurable as a passing phase, from which the earliest escape is desirable. Depreciated paper currencies have accordingly a tendency to right themselves, and already in the Argentine Republic, considerable progress has been made in this direction. We may regard depreciated currencies as passing ailments, more or less serious, which carry, so to speak, their own remedy.

The change in the value of silver as measured in gold is of a different character. It is not due to any illegitimate issue of promises to pay, in the form of paper notes. It is not a change of which any country need be in the least ashamed, for silver money is just as much "hard" money, as is gold money. The change to a lower value for silver as measured in gold, is not a passing phase but one more or less permanent, owing to the vast decrease in the demand for a metal of which the supply is, taking one year with another, wonderfully constant. The change accordingly, to a fluctuating relative value between gold and silver, must be absolutely permanent, unless and until governments revive the practice of a fixed ratio. Uniformity of price between gold and silver was—in the words of Mr. Bagehot—"generated by the practice of governments; it ceased when the practice ceased, and *will not revive till the practice revives.*"

XXXII

SUMMARY

IN the foregoing pages we have depicted the history of the joint standard, and the present phase of the controversy as to its re-establishment. We have seen how the joint standard provided the world with a common currency down to 1873. How, as it were by accident, owing to the want of knowledge and of prescience on the part of Lord Liverpool, the first inroad upon the bimetallic system was made. How the result of the change he inaugurated was long concealed by the operations of the Latin Union, and how finally, in 1873, partly in pique and partly no doubt in real fear of possible consequences, France ceased to coin silver freely at her mint, and the fixed ratio was abandoned. Since then the western world has been without a stable currency, and the world as a whole has lost its common currency, from whence innumerable inconveniences and disasters have resulted, though, as it is an ill wind that blows nobody good, some have been able to fish successfully in the troubled waters.

The result of the loss of the joint standard is, that we have on the one hand a group of states trading with a gold standard of currency only, and we have on the other hand another group of states, embracing a far larger aggregate population, where the wages of labour vary from a penny to fourpence a day, trading with a silver standard of currency only.

It is the desire of bimetallists to create a league of states, which shall by agreement and simultaneous action undertake, to freely coin both gold and silver as legal tender at a fixed ratio. Past experience has proved that when such a ratio is established, even by relatively a few states, it governs the market price of silver and gold, so that the value of either metal can only diverge from it, to an extent sufficient to pay the cost of importing or exporting specie, as the course of trade may demand. No attempt to demonetise gold has been made, nor is such an attempt even in contemplation, although the Gold Standard Defence Association, perhaps from a sense of the natural fitness of things, has in its title embodied this fallacy : for its name implies that the gold standard is in peril, and therefore in need of defence. Bimetallists, however, have no desire to abandon the gold standard, on the contrary they desire to make it universal instead of local, and this can only be done by linking silver with gold, whereby gold will once more be current everywhere.

It was the opinion of Mr. Bagehot that if France had adhered to the fixed ratio, the effect of the mutation by Germany of her currency from silver to gold, would have been "rendered scarcely observable." But we cannot blame France for exhibiting restiveness under the circumstances, more especially as we were originally to blame in discarding the joint standard in 1816. The result of the change in her currency laws by France, has been the break in the relative value of the two metals. It is not due to any change in the relative output of the two metals. In the words of Mr. Bagehot, "The depreciation of silver as against gold is not caused by a sudden excess in the supply of silver, for the new supplies of silver have been only moderate, and none of them have come here." The depreciation has been caused by legislative enactments. The fall of silver 25 per cent in value within fourteen days after the closing of the Indian mints, is a fact which stands out in such clear relief that no casuistry can explain away.

In addition to the advantage to the world of recovering a common currency, we should have in the joint standard a currency more stable in value than a currency based on a single metal can possibly be. Some inequalities of output of one metal there must always be, but when the two are linked together, these inequalities tend to equalise each other, resulting in greater uniformity of production and greater steadiness of value in money. We have also pointed out, that although the purpose of bimetallism is to restore to the world a stable common currency, there is every indication that the restoration of such a common currency would lead to improved trade, for when silver approximated more nearly to its old value for a time in 1890, there was a marked revival in trade, and again, when early in 1895, silver began to rise in gold value, trade immediately responded.

Let us bear in mind the accidental way in which the joint standard of currency came to be abandoned, first in England, and then in France and the United

States. And let us further bear in mind, that practical men of world-wide commercial experience have never condemned the joint standard. Mr. Alexander Baring (Lord Ashburton) in 1828 said, "I have always thought that it is possible and desirable to maintain in this country a silver currency as legal tender, founded on the proportion of silver to gold established in the currency of France. I have no doubt of it." In 1869 Baron Alphonse de Rothschild said, "Had I to choose a system with the experience we have, I should not hesitate to accept that of the double standard."

The real strength of the monometallic position lies not in argument, but in the fact that the monometallists are in possession. It is the fate of all originality of thought to be at first opposed. Prudence cautiously avoids what looks like a new path. But when once perseverance is rewarded, it is soon asserted that the whole matter was so obvious, that the only cause for astonishment was that it ever escaped observation. In the words of Malebranche, "It is not thought fit to disturb common opinion, for it is not truth which influences the world as it exists, so much as opinion and custom."

APPENDIX

Mr. SAUERBECK'S INDEX NUMBERS

(To which Mr. Arnold Hepburn has added the annual average price of Silver in London.)

Year.	Index-number of 45 principal Commodities.	Index-number of Silver. 100 = 60.84d.	Annual average price of Silver in London.	Year.	Index-number of 45 principal Commodities.	Index-number of Silver. 100 = 60.84d.	Annual average price of Silver in London.
1820	112	...	...	1858	91	101.0	$61\frac{7}{16}$
1821	106	...	...	1859	94	102.0	$62\frac{1}{16}$
1822	100	...	...	1860	99	101.4	$61\frac{1}{8}$
1823	103	...	...	1861	98	99.9	$60\frac{3}{4}$
1824	106	...	...	1862	101	100.9	$61\frac{3}{4}$
1825	117	...	...	1863	103	101.1	$61\frac{1}{2}$
1826	100	...	...	1864	105	100.9	$61\frac{3}{4}$
1827	97	...	...	1865	101	100.3	61
1828	97	...	...	1866	102	100.5	$61\frac{1}{8}$
1829	93	...	...	1867	100	99.7	$60\frac{9}{16}$
1830	91	...	...	1868	99	99.6	$60\frac{1}{2}$
1831	92	...	...	1869	98	99.6	$60\frac{1}{2}$
1832	89	...	...	1870	96	99.6	$60\frac{1}{2}$
1833	91	97.2	$59\frac{3}{16}$	1871	100	99.7	$60\frac{9}{16}$
1834	90	98.4	$59\frac{13}{16}$	1872	109	99.2	$60\frac{1}{4}$
1835	92	98.1	$59\frac{11}{16}$	1873	111	97.4	$59\frac{1}{4}$
1836	92	98.9	60	1874	102	95.8	$58\frac{1}{16}$
1837	94	97.7	$59\frac{7}{16}$	1875	96	93.3	$56\frac{3}{4}$
1838	99	97.7	$59\frac{7}{16}$	1876	95	86.7	$52\frac{3}{4}$
1839	103	99.5	$60\frac{3}{8}$	1877	94	90.2	$54\frac{7}{8}$
1840	103	99.5	$60\frac{3}{8}$	1878	87	86.4	$52\frac{9}{16}$
1841	100	98.7	$60\frac{1}{16}$	1879	83	84.2	$51\frac{1}{4}$
1842	91	97.6	$59\frac{7}{16}$	1880	88	85.9	$52\frac{1}{4}$
1843	83	97.2	$59\frac{3}{16}$	1881	85	85.0	$51\frac{11}{16}$
1844	84	97.1	$59\frac{1}{8}$	1882	84	84.9	$51\frac{5}{8}$
1845	87	97.4	$59\frac{1}{4}$	1883	82	83.1	$50\frac{9}{16}$
1846	89	97.5	$59\frac{5}{16}$	1884	76	83.3	$50\frac{11}{16}$
1847	95	98.1	$59\frac{11}{16}$	1885	72	79.9	$48\frac{5}{8}$
1848	78	97.8	$59\frac{1}{2}$	1886	69	74.3	$45\frac{3}{8}$
1849	74	98.2	$59\frac{3}{4}$	1887	68	73.0	$44\frac{5}{8}$
1850	77	98.7	$60\frac{1}{16}$	1888	70	70.4	$42\frac{7}{8}$
1851	75	99.9	$50\frac{13}{16}$	1889	72	70.2	$42\frac{11}{16}$
1852	78	99.9	$60\frac{13}{16}$	1890	72	78.2	$47\frac{11}{16}$
1853	95	101.2	$61\frac{9}{16}$	1891	72	74.1	$45\frac{1}{16}$
1854	102	101.1	$61\frac{1}{2}$	1892	68	65.4	$39\frac{13}{16}$
1855	101	100.7	$61\frac{1}{4}$	1893	68	58.6	$35\frac{5}{8}$
1856	101	101.0	$61\frac{7}{16}$	1894	63	47.6	$28\frac{7}{8}$
1857	105	101.5	$61\frac{3}{4}$	1895	62	49.1	$29\frac{3}{4}$

COMMERCIAL RATIO OF SILVER TO GOLD EACH YEAR FROM 1687 TO 1895, BASED UPON THE PRICE OF SILVER IN THE LONDON MARKET.

Compiled for the Bimetallic League, 29 Cornhill, E.C.

(*From* 1687 *to* 1832 *the ratios are taken from Dr. A. Soetbeer; from* 1833 *to* 1878 *from Pixley and Abell's tables; and from* 1879 *to* 1895 *from daily cablegrams from London to the Bureau of the United States Mint.*)

Year.	Ratio.	Year.	Ratio.	Year.	Ratio.	Year.	Ratio.	Year.	Ratio.
1687	14.94	1729	14.92	1771	14.66	1813	16.25	1855	15.38
1688	14.94	1730	14.81	1772	14.52	1814	15.04	1856	15.38
1689	15.02	1731	14.94	1773	14.62	1815	15.26	1857	15.27
1690	15.02	1732	15.09	1774	14.62	1816	15.28	1858	15.38
1691	14.98	1733	15.18	1775	14.72	1817	15.11	1859	15.19
1692	14.92	1734	15.39	1776	14.55	1818	15.35	1860	15.29
1693	14.83	1735	15.41	1777	14.54	1819	15.33	1861	15.50
1694	14.87	1736	15.18	1778	14.68	1820	15.62	1862	15.35
1695	15.02	1737	15.02	1779	14.80	1821	15.95	1863	15.37
1696	15.00	1738	14.91	1780	14.72	1822	15.80	1864	15.37
1697	15.20	1739	14.91	1781	14.78	1823	15.84	1865	15.44
1698	15.07	1740	14.94	1782	14.42	1824	15.82	1866	15.43
1699	14.94	1741	14.92	1783	14.48	1825	15.70	1867	15.57
1700	14.81	1742	14.85	1784	14.70	1826	15.76	1868	15.59
1701	15.07	1743	14.85	1785	14.92	1827	15.74	1869	15.60
1702	15.52	1744	14.87	1786	14.96	1828	15.78	1870	15.57
1703	15.17	1745	14.98	1787	14.92	1829	15.78	1871	15.57
1704	15.22	1746	15.13	1788	14.65	1830	15.82	1872	15.63
1705	15.11	1747	15.26	1789	14.75	1831	15.72	1873	15.9
1706	15.27	1748	15.11	1790	15.04	1832	15.73	1874	16.17
1707	15.44	1749	14.80	1791	15.05	1833	15.93	1875	16.59
1708	15.41	1750	14.55	1792	15.17	1834	15.73	1876	17.88
1709	15.31	1751	14.39	1793	15.00	1835	15.80	1877	17.22
1710	15.22	1752	14.54	1794	15.37	1836	15.72	1878	17.94
1711	15.29	1753	14.54	1795	15.55	1837	15.83	1879	18.40
1712	15.31	1754	14.48	1796	15.65	1838	15.85	1880	18.05
1713	15.24	1755	14.68	1797	15.41	1839	15.62	1881	18.16
1714	15.13	1756	14.94	1798	15.59	1840	15.62	1882	18.19
1715	15.11	1757	14.87	1799	15.74	1841	15.70	1883	18.64
1716	15.09	1758	14.85	1800	15.68	1842	15.87	1884	18.57
1717	15.13	1759	14.15	1801	15.46	1843	15.93	1885	19.41
1718	15.11	1760	14.14	1802	15.26	1844	15.85	1886	20.78
1719	15.09	1761	14.54	1803	15.41	1845	15.92	1887	21.13
1720	15.04	1762	15.27	1804	15.41	1846	15.90	1888	21.99
1721	15.05	1763	14.99	1805	15.79	1847	15.80	1889	22.10
1722	15.17	1764	14.70	1806	15.52	1848	15.85	1890	19.76
1723	15.20	1765	14.83	1807	15.43	1849	15.78	1891	20.92
1724	15.11	1766	14.80	1808	16.08	1850	15.70	1892	23.72
1725	15.11	1767	14.85	1809	15.96	1851	15.46	1893	26.49
1726	15.15	1768	14.80	1810	15.77	1852	15.59	1894	32.56
1727	15.24	1769	14.72	1811	15.53	1853	15.33	1895	31.26
1728	15.11	1770	14.62	1812	16.11	1854	15.33		

(1873–1895: DEMONETISATION)

INDEX

AGRICULTURAL labour displaced, 80
Agriculture, 102
Aldenham, Lord, quoted, 33, 126, 139
Alison, quoted, 7
Allard, M., quoted, 132
Altering the standard, 111
Arable land laid down, 80
Argentine Republic and gold, 122, 134, 142
Ashburton, Lord, 149
Australian output of gold, 22, 25, 109, 138

BAGEHOT, W., quoted, 18, 24, 145, 147, 148
Balfour, Mr., quoted, 65
Bank of France, bimetallist, 15, 16
Bank deposits, 66
Bankers and gold, 10, 65
Bankers' Magazine, quoted, 127
Baring, Alexander, quoted, 149
troubles, 121, 143
Barter, all trade is, 45
Beaconsfield, Lord, quoted, 8, 16
Belgium, 12
Bimetallism, 3
old and new, 114
national and international, 117
why abandoned, 5, 15
Bland Act, 54, 115
Bordeaux, exchange disturbed at, 15
Bounties on exports, 103
Bristol Chamber of Commerce, 125
British Association, 76
Brussels Monetary Conference, 106
Bryan, Mr., 55, 56

CAIRNES, Professor, 25
California, output of gold, 22, 25, 109, 138
Child, Sir Josiah, quoted, 85
Chili, 140
Clipping the currency, 5
Cobden Club, the, 41
Comstock Silver Mine, 33
Consols in 1748, 7
price of, 66
Consumption of foods, 82
Contraction of currency, 43, 135
Corn Laws, 26
Cornish tin mines, 80
Cotton-spinning industry, 80, 81
Courtney, L., quoted, 75
Crown-piece, the, 62
Cubic feet of gold and silver, 61
Currency depreciated, 78

DEBTORS and appreciated gold, 69, 91
Demonetisation of silver, cloaked, 12, 14
by Latin Union, 61
in the States, 51
Depreciating the currency, 78
Depression of trade, 135
Dishonest money, 56, 59, 63, 92
Distribution, theory of better, 35
Dumping-ground fallacy, 117

Economist, The, 30
Elizabeth, Queen, 57
Employers and low prices, 77
England, forced paper issue in, 13
prosperity of, 140
Erasmus, quoted, 4

Exchanges disturbed, 15
Experiment, bimetallism an, 113
Exports of United Kingdom, 40, 110

FAIR TRADE, 104
Falling prices, 71, 120
Farrer, Lord, quoted, 59, 60, 83, 91, 94, 111
Four qualities of good money, 63
Foxwell, Professor, quoted, 83, 84, 115
France and token money, 11
  and the Latin Union, 14
Free trade, 42, 102
French ratio, 9
  Currency Act, 17, 18
  indemnity, 16

GAINS and losses fallacy, 131
Germany, 29
Giffen, Sir Robert, quoted, 40, 79, 84
Gold and the arts, 126
  as a crop, 110
  as natural currency, 11
  the cause of prosperity, 137
  cubit feet of, 61
  ounce of, 6
  output of, 32, 125
  price of, 26
  Standard Defence Association, 12, 147
  tons of, 33
  why prevalent in England, 9
Graham, Sir James, 74
Grass land laid down, 80
Great Broken Hill Mine, 34
Greece, 12
Green's *History*, quoted, 139
Gresham law, 57
Grocer's son, illustration, 131
Guineas, 6

HAMBURG Exchange disturbed, 15
  Chamber of Commerce, 17
Hard money, 145
Harvey, Mr., quoted, 46
Hats, illustration, 94
Hicks-Beach, Sir M., 25
Hiogo, 81
Hobgoblin fallacies, 114
Honest money, 56, 59, 63, 92
Hume, David, quoted, 73

IMPORTS of United Kingdom, 40, 110
Income-tax returns, 88
Index-numbers, 28
India, 23, 45
  and home payments, 68
  and wheat, 143
Indian mints closed, 93
  mutiny, 87
Inflation of currency, 133
Irish famine, 42
Italy, 12

JEVONS, Professor, 23, 73, 84
Joint standard, 6, 14

LABOUR, 72, 78
  cheap, 147
  leaders, 84
Latin Union, the, 12, 18, 23, 27
Letters, increase of post, 87
Liverpool, Lord (Prime Minister), 8, 11
  his errors, 12, 13, 14, 17, 141, 146
Locke, John, 6, 21, 51, 114
Losers by joint standard, 130
Losses and gains, 131
Low prices no evil, 71

MACAULAY, Lord, quoted, 5
M'Culloch, quoted, 73
M'Kinley, Mr., 55, 56
Mahany, Hon. R. B., quoted, 123
Malebranche, quoted, 149
Manchester Chamber of Commerce, 80
Markets, falling and rising, 120
Measures and weights fallacy, 74
Mill, J. S., quoted, 74, 84
Mississippi Valley, 30
Molesworth, Sir G., tables of, 38, 106
Mongredien, Mr., quoted, 40
Monk, General, quoted, 135
Monometallism, 3
  a question-begging term, 2

NAPOLEON III., monometallist, 15
National debt, 60
Nature does not produce money, 20
Newmarch, W., quoted, 83, 96

INDEX

Newton, Sir Isaac, 6, 21, 51, 114
  recommended identical ratio, 7
Nomos and money, 20
Norway, 140

OUNCE of gold in money, 6

PAPER currencies, 141
Peel, Sir Robert, 13, 27
Pendulum, illustration, 119
Pennies, 27
Peru, 140
Playfair, Lord, quoted, 35
Police efficiency, 88
Portugal, 140
Post Office, 87
  Savings Bank, 66, 87
Postal convention, 118
Potato famine, 42
Pound sterling, the, 27
Power, Mr., quoted, 30
Price of gold, 26
Prices rising and falling unequally, 25
  rising, 83
Protection and bimetallism, 102, 103

QUALITIES of good money, 63

RAILWAY earnings, 39, 99
  mileage, 36
Ratio, a fixed, 20, 106
  altered, 9
  effect of differing, 48, 52
  English, in 1696, 6, 21, 23
  French, 9
  inequitable, 81
  proposed, 100
  wanted a, 91, 96
Reduction of wages, 79
Resumption of specie payments, 123
Rising prices, 71, 120
Roman Empire, fall of, 135
Rothschild, Baron Alphonse de, 15, 116, 149
Royal Commission of 1886, 23
Russian War, 86

SAUERBECK'S tables, 28, 36, 43, 83, 84, 138
  in full, 151
Savings Bank returns, 86

Schools, elementary, 87
Seyd, Ernest, 116
Sherman Act, 54, 92, 114
Shilling, the silver, 62
Shipping, tonnage of, 37
Silver, price of, 152
  cubic feet of, 61
  demonetised, 8, 12, 14, 19
  legal tender to 40s., 7
  output of, 31, 91
  stability of, 29
Single standard of money, 128
Smart, Dr., quoted, 76
Smith, Adam, and small silver change, 10
  and money, 45
  and the ratio, 51
  and wheat, 135
Soetbeer, his views, 17
Soft money, 142
Somers, Lord, 6, 21
Specie payments, 123
Statist, The, 30
Statistical Society, the, 79
Stock markets fall, 127
Straits Settlements and tin, 80
Süess, Dr., quoted, 112
Sugar, consumption of, 82
Sweden, 140
Switzerland, 12

TENANCY, break in, illustration, 95
Tin mines, 80
  kettle, illustration, 99
Tobacco, consumption of, 82
Token money, 10
Trade is barter, 45
Transport theory, 35

UNITED STATES currency, 47
  token money, 10

VAN HELMONT, quoted, 124

WAGE-EARNERS, 72
Wages, reduction of, 79
War of Austrian Succession, 7
Weights and measures fallacy, 74
Wheat, price of, 135, 143
Widow, the, illustration, 98
William III. and the currency, 5
Wolowski, 116

www.ingramcontent.com/pod-product-compliance
Lightning Source LLC
Chambersburg PA
CBHW030245170426
43202CB00009B/638